Beer Terroir

BEER
TERROIR

*PLACE AND TASTE IN THE PRE-PROHIBITION
ROCKY MOUNTAIN WEST*

Braden Neihart

equinox

SHEFFIELD UK   BRISTOL CT

Published by Equinox Publishing Ltd.

UK: Office 415, The Workstation, 15 Paternoster Row, Sheffield, South Yorkshire S1 2BX

USA: ISD, 70 Enterprise Drive, Bristol, CT 06010

www.equinoxpub.com

First published 2025

British Library Cataloguing-in-Publication Data

A catalogue record for this book is available from the British Library.

ISBN-13 978 1 80050 690 9   (hardback)
        978 1 80050 691 6   (paperback)
        978 1 80050 692 3   (ePDF)
        978 1 80050 716 6   (ePub)

Library of Congress Cataloging-in-Publication Data

Names: Neihart, Braden author
Title: Beer terroir : place and taste in the pre-prohibition Rocky Mountain west / Braden Neihart.
Description: Sheffield, South Yorkshire ; Bristol, CT : Equinox Publishing Ltd, 2025. | Includes
    bibliographical references and index. | Summary: "Beer Terroir seeks to understand how the
    application of the concept of terroir-a term typically applied to wines-to beer can help us further
    our understanding not only of the beverage but also of the humans that occupy regions. The
    book is aimed at those interested in the history of beer and brewing and in The Rocky Moun-
    tains as a region, as well as in the culture and history of foodways"-- Provided by publisher.
Identifiers: LCCN 2025020262 (print) | LCCN 2025020263 (ebook) | ISBN 9781800506909
    hardback | ISBN 9781800506916 paperback | ISBN 9781800506923 ePDF | ISBN
    9781800507166 ePub
Subjects: LCSH: Beer--Mountain States--History--19th century | Terroir--Mountain
    States--History--19th century | Brewing--Mountain States--History--19th century |
    Breweries--Mountain States--History--19th century | Prohibition--United States--History
Classification: LCC TP573.U6 N45 2025 (print) | LCC TP573.U6 (ebook) | DDC
    663/.42097809034--dc23/eng/20250729
LC record available at https://lccn.loc.gov/2025020262
LC ebook record available at https://lccn.loc.gov/2025020263

Typeset by Scribe Inc.

Contents

Acknowledgments

When I began my grad school applications, I was unsure *what* aspect of history I wanted to research but just knew I wanted to read and write more history. Naturally, after I had submitted my hopeful (though likely vague) applications, I discovered the topic I wanted to pursue. In the University of Denver's library was Max Nelson's *The Barbarian's Beverage*. For the first time, I discovered beer history as a true and legitimate academic subject. There are many inspirations and contributors who have helped indirectly and explicitly, but I wanted my first thank-you to be to Nelson, for opening up a world of history and driving the last half decade of research.

My second thank-you is to my undergrad and graduate school advisers. Dr. Joyce Goodfriend was always encouraging and pushing me to "find a better word." I hope that I have some good ones here. Dr. Jared Orsi taught me to write in a way that I did not think possible and encouraged my beer history projects even when it might have made more sense to focus on one at a time. I hope this work is less error ridden than previous works, though it pairs equally well with a cold brew.

Additionally, I appreciate all the people who have listened, offered feedback, or read sections of this work. Explaining what I mean by beer terroir has been instructive in understanding how this term could operate better and clarified how I can explore this concept in new and interesting ways. I appreciate the comments and feedback, in particular, from Hayden Nelson and Allyson Brantley, who read essentially the entirety of this work and offered really crucial insights and ideas for clarity and expansion.

I would also like to thank all the bar space I have taken up with my laptop and books. I am fortunate to write about subjects I get to experience in person and have found writing in breweries immensely rewarding. Drinking beer in a brewery surrounded by people sharing their lives over beer was a constant and helpful reminder of the continuity of the meanings of beer. It also provided many opportunities to draw inspiration from beer terroir advertised in modern Rocky Mountain West breweries. Many beers have gone into the production of this book, and I am appreciative of the ones I have consumed and the ones consumed within these pages.

I began this work as an independent scholar and was without university affiliations. Over the course of this book, I worked first as an adjunct and finished as a faculty member at the Community College of Aurora. I am thankful for the access my community college provides to online databases and resources. It provided a solid start and allowed me to expand my research avenues further than I could before. To complement this, I relied heavily on online archives and digitized sources. I would like to thank all the online archives and the archivists who work constantly to maintain, update, and expand them. In particular, I would like to thank all the archivists involved in creating and maintaining Colorado Historic Newspapers, History Colorado, and the Denver Public Library. Without these resources, simply put, this project would have been impossible. I want to thank the many workers who scan and upload newspaper documents, journals, diaries, images, and all sorts of invaluable sources for historians.

Finally, I would like to thank my favorite beer buddy—my wife. She has heard more complaining and brainstorming about this project than anyone ought to endure, all the while encouraging me to finish it and stay on task. I would like to dedicate this book to her for always supporting me, listening to my ideas, and helping me organize my half-baked ideas into coherent sentences.

Preface

Alcohol has long been a part of the human experience. Today, there is a tremendous variety in the types of alcohol available to consumers. Various grain or plant bases, yeasts, alcohol content, fermentation practices, and so forth present a dazzling array of options for alcohol consumers. With all this variety, one would imagine many alcohol drinkers are well suited to taste and determine which alcohols they prefer, what they want, and how they want to consume them. This is not true.

Researchers and psychologists have conducted a number of studies to test how well humans can identify the alcohol they are drinking. A 2021 study conducted in Budapest led by Vivien Bodnár used rum and a placebo rum with the same aromas with a group of over one hundred participants. They wanted to know three things: Will people take on inebriated behavior if they are drinking nonalcoholic placebos but are told they are drinking alcohol? How does this translate to a social environment? Can people who are drinking alcohol or nonalcoholic drinks be "convinced" they are sober or drunk when they are, in fact, the opposite?[1]

What they found in their study was that people in social settings are less influenced by a perception of drunkenness or sobriety. They know, when interacting with others, if they are drinking intoxicants or not. Yet when alone, people were unable to identify if they were consuming alcohol or a placebo.[2]

This does not just apply to hard spirits and liquors. The rise in non-alcoholic (NA) beer has opened up doors for researchers to examine how adept people are at understanding what they are consuming. Paul Smeets and Cees de Graaf led an investigation to understand brain reward systems when consuming alcohol or NA beer. What they discovered is while participants were able to tell the differences between tastes of different beers, they were unable to identify which contained alcohol or not.[3]

Humans have consumed alcohol for millennia, and it is fascinating that we cannot tell, between two similar drinks, which contains alcohol. Exploring taste and perception is often in the domain of psychology and natural sciences. This study applies the study of taste perception to historical settings. Although we will never be able to perfectly understand what something produced a century ago tasted like, we can try to understand the ways that people influenced others' tastes and perceptions. Using the framework of terroir, applied to beer and layered on the American pre-Prohibition Rocky Mountain West, we can understand the ways that people have always attempted to influence taste. Indeed, as research shows, humans are incredibly gullible in what they do or do not taste, and thus there is fertile ground for marketers to step in and inform, educate, or cajole consumers into tasting preferred flavors.

Why does examining beer terroir of any time or place matter? Ultimately, I maintain it comes down to three human reasons. First, one's enjoyment of a beer does not depend on their understanding of terroir. Yet cognition of terroir and taste heightens the drinking experience. Wine snobs may be irritating at dinner parties, but it is unquestionable they extract far more pleasure out of the same glass of wine as others. Having an understanding or background with which to taste and consume beer has the potential to increase one's appreciation of beer.

Second, the craft beer revolution since the 1980s has attempted to elevate the status and quality of beer. This is particularly true in the US, where beer has been monopolized by a few major corporations since the end of Prohibition. This work, same as others in beer history, is an attempt to raise general awareness of beer's complexity. As will hopefully be shown below, it is a complicated drink with a fascinating past and present—something that often gets lost in the general alcohol culture.

Third, relatedly, beer is an immensely common drink. Super Bowl Sunday is notorious for alcohol consumption. In 2021, during the middle of the COVID-19 pandemic, beer sales for Super Bowl Sunday reached $1.3 billion.[4] We consume beer at rapid rates throughout the year and throughout our weeks constantly. Beer is an everyday drink for many, which I think warrants a moment of pause and appreciation of its terroir and taste. Refusing to—or at least attempting to refuse to—take a beverage consumed by millions of Americans daily for granted may help us understand and appreciate the complex food patterns and production systems that provide us with our calories and nutrients.

Economics are always in the minds of brewers and publicans. Beer is a cultural product, but at the end of the day, it was created to sell. Many of the aspects of terroir put forth by the brewing industry were indeed an act to sell more beer. Yet they were building something in doing so. In their efforts to boost products, they formulated and constructed the beer terroir of the Rocky Mountain region and laid foundations that continue to influence, inspire, and instruct brewers there today. This is the heart of this work, the unintentional and communal efforts taken by individuals that culminated in a framework encompassing a subregion of the North American continent. Built by major brewers and beer industry magnates as much as by bartenders and laborers, the terroir of beer contains multitudes.

Chapter 1 outlines the framework and lens through which we can understand the connection between terroir and taste. What we taste in beer or other edible substances is a complex mix of the environmental forces that produced that product. Another layer that elevates certain notes and diminishes others is what we are told to taste. Marketing, promotion, labels, and so forth are all strategies to enable consumers to pick up on desired flavors.[5] Our taste is flexible as we are influenced by a range of factors.

The second chapter explores the geography and environmental aspects of the Rocky Mountain West. Theoretical frameworks are only as valuable as the reality they reflect. The Rockies are a particular landscape that differs substantially from French-vineyard-derived visions of terroir. This chapter argues the value of terroir in application to beer as it helps us read the Rocky Mountain West political, cultural, and foodway landscape differently. The Rockies are a distinct landscape and therefore have a commensurate beer terroir.

Continuing this thread, chapter 3 engages in the pursuit of water in the Rockies. This action involves both the acquisition of sufficient water to brew in addition to the ongoing debates in the nineteenth century over water purity. Water purity extends beyond the nutritional impact on the body. It also contains essential ingredients for pure beer. Growing Prohibition sentiments pushed back against water usage for alcohol production; brewers countered with health concerns about water. Water purity continues to feature centrally in Rocky Mountain brewing, distilling, and bottling. The rhetorical and physical work brewers performed to claim pure water for brewing was remarkable and illustrated the plasticity of terroir. Brewers fought over water, arguing at times opposing sides of water safety.

We cannot actually taste history, but our ideas of history and culture inform our senses. Chapter 4 investigates this phenomenon. How white brewers, in their newfound home, re-created traditional Euro-American brewing practices was a challenge, but one they eagerly sought to overcome. When miners and their affiliated colonists moved into the Rockies, they imported ideas about foodways and beer. As they transformed the environment into a new way to extract wealth, brewers found ways in turn to build familiar spaces for miners and others. Beer history informed the methods used to produce beer, but their environs posed challenges to this in practice.

Chapter 5 begins to deal with the brass tacks of brewing. Beyond water (discussed in chapter 3), chapters 5 and 6 address the ingredients essential to brewing. Barley, hops, and additives position brewers in the Rockies in a difficult situation. Few of these ingredients are produced locally, forcing them to search farther afield for them. Beyond this, they make moral choices to use ingredients that are not in line with their traditional brewing practices but that economics and geography necessitate.

Next, chapter 7 refines the sense of place and the provincialism embodied by local brewers in the Rockies. In more recent years, there has been a movement toward local consumption of food and drink spurred by concerns of climate change, freshness, and supporting the local economy. What is considered local, however, may include a much wider sphere today than in the past. Before Prohibition, brewers in the Rocky Mountain West, as elsewhere, maintained a tight focus on their local community for production and consumption. Local beer was not just beer from the region or the state but often argued only to be beer produced in a single town or community. This emphasis arose in part to protect budding breweries from larger out-of-state operations or even those in larger metropolises within the region.

Thus far, much has been said about the efforts of brewers to use strategies to set their beer apart in ways that can be applied to beer terroir. Chapter 8 investigates the successes or failures of these strategies. Ultimately, one can say their beer is better than others' due to its tangible and intangible qualities, but does it sell? Do people believe and buy it in sufficient quantities to justify the claims made by brewers, and what competing and countervailing forces complicate success and failure in this environment?

Not every aspect of beer and brewing is part of terroir. Though a spacious term, chapter 9 explores the ways that in the pre-Prohibition Rocky Mountain West, brewmasters and those advertising their wares did not consider every step of the brewing process worthy of mention. The day laborers, major investments, and equipment are some of the aspects that certainly could be considered part of terroir but were noticeably absent from the advertisements and other materials produced. This does not necessarily mean that those in charge did not consider them important—in fact, it could clearly be seen as the opposite. However, in terms of what set one beer apart and ahead of its competitors, there are clear steps or components of brewing that were left out.

Terroir contains many elements, and one of the more powerful ones is the impact of memory and nostalgia. Chapter 10 addresses the current nostalgic state of brewing and beer in the Rocky Mountain West—in particular, the ways that stories, legends, and the history of the American West continue to influence brewing through naming conventions, styles, and marketing strategies. Breweries today tap into the terroir built by brewers from a century ago and play to consumers' imaginations of the American West and its association with gold rushes, cowboys, and frontier living. Here we explore the ways that nostalgia for the history covered in the preceding chapters informs our taste today. Brewers in the nineteenth-century Rocky Mountains were seeking to brew beer in a way that was reminiscent of their homes along both sides of the Atlantic Ocean. Today, brewers and advertisers incorporate many of the same ideas. The difference, however, is how they pull from the Rockies' history to inform our taste buds.

In closing, chapter 11 looks at the ways that we can continue the conversation about beer terroir and expand our historical and contemporary understanding of our relationship to beer and brewing, especially in terms of our physical and emotional connection to certain places—either regions or specific establishments—that foster relationships.

Ultimately, this work seeks to expand the value of beer as a lens into the past. There is much that connects humans across time and space, and our enjoyment and pursuit of familiar intoxicants allow us today to understand past motivations and goals. In particular, thinking about beer landscapes and beer terroir allows historians to combine subjects and issues into a single framework that often operates in separate spheres. I believe it is fun, enjoyable, and productive to think about the past through beer.

Introduction

In 1713, on the 19th of May, Robert Livingston—a preeminent colonial New York merchant and politician and grandfather of the Founding Father Robert Ross Livingston—received a letter from Johanna Van Horne Livingston. Bearing bad news, it reported a barrel of spoiled beer, reportedly due to a loose bunghole that allowed air to seep in, oxidize the beer, and thus ruin the barrel.[1] Unfortunately, such occurrences were common before widespread refrigerated transportation and stainless-steel kegs.

Even the slightest allowance of oxygen portended disaster for beer. Joanna had requested this beer a few days before and was upset that Robert had not sent it sooner.[2] Having received a delayed order of beer partially empty from leakage and spoiled from heat, humidity, and oxygen did not put Robert in her good graces. Robert was typically a trustworthy source for shipping beer, as he had sold some to the Earl of Bellomont in 1699 and delivered it to troops in Albany six years earlier in 1693.[3] This time, however, physics and chemistry conspired against him and ruined the beer he belatedly delivered.

Nearly two hundred years later and half a continent away, brewers and consumers contended with similar issues of spoilage via transport. Begun in 1890 by three members of the Becker family, the Becker Brewing and Malting Company operated in Ogden, Utah, to meet a dire need for quality beer. Run by a father—an immigrant from Germany—and his two sons, this brewery's mission was to "produce a beer of extra fine quality and purity" with an added goal of making beer "better than that produced and shipped in from the eastern states."[4]

By and large, the Becker family succeeded in their goal of producing fine, local beer. In the first decade of the nineteenth century, Becker beer was the purported preference of all retailers in Ogden. More than that, Ogden's medical community prescribed "it in preference to any other brand, knowing it to be absolutely pure and free from all deleterious preservatives."[5] Their quality beer led the Beckers to substantially increase their production. Over twenty years, their consumption of barley jumped from five thousand bushels a year to over thirty thousand. Much of this barley came from Utah farms and was happily reported to be "the best barley grown in the world."[6] Likewise, hop consumption poundage grew from four thousand to fifty thousand per year.

This, in a nutshell, is the story told below. Miners crisscrossed the Rockies in search of gold with a significant number of auxiliaries hoping to strike it rich off those who worked in the mines. As these populations moved and settled—all dependent on the success of the mines and the capacity for economic development and diversification—they wanted foodways that sustained their efforts. How did white people move into the Rocky Mountain region and establish brewing operations reminiscent of those back east—which could mean as close as St. Louis or as far as Germany? How did brewers embrace their new climates, altitudes, and other features and turn them to their advantage over more established, profitable, and powerful breweries back east?

Beer was one such foodway that connected people to their traditional foodways while they toiled in a new environment. As brewers established businesses and marketed their products, they established a communal beer terroir of the pre-Prohibition Rockies. Their adaptation to their new homes was detailed in their calls to drink locally produced beer. Superior to imported swill, brewers quickly incorporated the geographical, environmental, and historical aspects of their new environs into reasons for their quality products. Indeed, there were benefits in the Rockies not present in the East. Yet more often, these attachments of terroir to beer were in response to challenges and difficulties. The process of transforming brewing obstacles to beer terroir is the core of this work.

Tasting Beer

Spoiled beer from heat, humidity, or other reasons can be immediately detected from smell and taste. Barrels that allow too much oxygen to enter, or sat in the sun too long, will result in far worse tasting beer than one that has been properly stored and cared for. These are realities that our taste buds warn us of. Drinking spoiled beer can lead to illness and suffering, and our olfactory and taste receptors reveal the potential harm before we consume harmful substances. These senses, however, can be tricked and manipulated. In effect, they can be led to taste certain flavors or directed away from noting foulness. Or, in some food products, our senses interpret spoilage or rottenness favorably. Taste buds are not purely objective.

Taste and our world are the construction of our perceptions. Tasting allows us to distinguish between foods that our bodies desire—such as juicy, fatty, meaty burgers—and those that will harm us—such as unripe fruit or poisonous plant matter. From our earliest days, humans as a species and as individuals have used taste to navigate our world. Even well beyond the infant stage of life, putting new objects into our mouths to discern their favorability is a fundamental human action. Yet for how crucial our taste buds are to us, taste is by no means a static, definitive metric. Although we may all say harmful substances taste bad, or foul-smelling food depletes our appetite, the world of edible and enjoyable foods can elicit substantial differences. What is enjoyable to one person may offend others' stomachs; this is true across cultures as much as between people within a family.

Our taste buds are one part of a vast network of brain synapses and neurological signals. These come not only from what touches our ten thousand taste buds but also from what we read, hear, or observe. Growing up, if our family rejects a type of food or way of cooking, our bodies may not appreciate the tastes it offers, as we have created a predisposition against that food. What we are told—implicitly or explicitly—to enjoy or disdain shapes our palettes.

One of the most powerful elements of taste is marketing and advertising. These messages come from family and friends as much as print or other advertisement sources. We detect what we are expecting to taste. These expectations come from figures of authority and a barrage of marketing building expectations of taste and enjoyment. What we are told to taste and to expect largely informs what we do indeed perceive in what we consume. Ads cannot convince us not to taste certain distinguishable notes, but they can direct us toward more yummy aspects. The stories advertisements tell and the images they conjure in our minds are powerful. When we decide to consume certain products based on their advertisements, in some capacity, we are seeking to place ourselves in those stories and take part in the narrative.

In terms of food and foodways, these narratives connect us to an ingrained understanding of what we consume. We care about not only healthy or harmful consumables but also the history and familiarity of diverse foodways. In every culture, foods are prepared for certain holidays or events. Beyond this, food is comforting and medicinal. When we are sick or in mental, emotional, or physical pain, we gravitate toward certain dishes. What foods remind you of home? Your childhood? What restaurants or meals do you seek out after a hard day? How do you celebrate milestones and accomplishments? Answering these questions asks us to think about the value food has in our lives. Beyond nutrition, food and drinks are a central part of our identity and how we navigate the world. Calories are cultural.

This work focuses on one aspect of foodways: beer. More specifically, it examines the importance of beer to a group of people in a new environment. Emigrants carried cultural valuations of beer with them and at the same time constructed new meanings. Of course, certain groups brought various brewing and drinking traditions. But on a larger scale, many European immigrants to the US transplanted a brewing culture. In this way, I want to center the beverage and examine the various values, meanings, and efforts contained within each bottle and sip.

To do so, I will examine the terroir of beer in the Rocky Mountain region from the mid-nineteenth century until Prohibition. Beer terroir allows us to examine the confluence of tangible and cultural elements of beer. It provides an avenue to understand the actions of people in a new place attempting to re-create familiar and safe spaces. Its usefulness in exploring differences between American beer regions adds a layer of nuance and detail often missing from American beer history. For the last several decades, American beer has been dominated by mass-produced beers that obfuscate the historical richness within American beer history. Beer terroir is one attempt to revive this history and enliven current beer culture in the US.

Terroir in this work is understood to be the flavor and taste profile imparted from the physical environment, cultural processes, societal values, and expectations of food products. It is the confluence of tastes based on chemical, cultural, and capitalist processes. Terroir is what we taste, but what we taste is the cumulation of climate, geography, ingredients, marketing, spoilage, choices, and unintended consequences. Beer terroir is the application of the above definition of terroir to beer and brewing.

This traces the rough bounds from the Colorado Gold Rush in 1859, which brought white people semipermanently to the Rocky Mountains. White people had crossed these mountains before on the way to the West Coast, and of course, Native Americans had crossed them for millennia. However, it was from 1859 to 1920 that white beer-brewing people settled in the Rocky Mountains and produced great quantities of beer. After this, people, of course, continued to live in this region, but beer production ended during the dry decade.

Each chapter will parse out one of the multitudinous aspects of terroir and examine how brewers, pundits, and the general populace approached and interacted with these categories. By no means an exhaustive list, this book instead aims to both introduce this manner of thinking about the history of beer's regional flavoring and prompt beer drinkers to consider the environmental, geographic, human, and cultural labor that goes into providing a glass of beer. Throughout, beer—its production, value, nutrition, history, and more—is the main character.

In order to best capture the essence of beer terroir, I want to make a note about source use. By and large, the creation of beer terroir in the nineteenth-century American Rockies was a public endeavor. It did not matter as much what people thought privately or wrote in journals. Instead, what carried the most significance was how people communicated about beer and brewing in public spaces. Newspapers were the public spaces where brewers, publicans, drinkers, boosters, and others made their cases and promoted beer terroir. Beer terroir was promoted, produced, and proliferated in this public domain. Brewers and town boosters from the same vicinity boasted the superior terroir of their locale. Communally, terroir is the product of this process repeated across the Rocky Mountain West. In this light, marketing schemes repeated and mimicked across the area take prominence. The methods used to promote beer in the nineteenth-century Rocky Mountains took on a regional flavor.

This appears quite plainly in their animosity toward anything deemed a foreign beer. A local man in Livingston, Montana—Mr. A. Landt— established the Apollo Springs Brewery in 1884. Landt believed that his establishment would "drive all foreign beers from his town."[7] That same year, in Las Vegas, New Mexico reported that Las Vegas beer would "soon drive foreign beers out of the market."[8] Of course, there were variations on this theme. Klein's Brewery in 1880s Carson City, Nevada, often received praise for "the superiority of Klein's beer over all others, foreign and domestic."[9] This is not to say that enmity toward nonlocal beer was unique to the West—merely that their marketing schemes resembled and echoed one another in such a way that reverberated through the region.

This is certainly not the first work to consider terroir or to consider the importance of place and taste. Wine has long had a history of terroir and has been the recipient of countless scientific, sociological, and historical studies examining the real and perceived impact of terroir on the final product. Not just aficionados or sommeliers, but many laypeople and imbibers are familiar with the narratives and discourses of wine terroir. Beer and brewing, in recent years, have experienced a number of historical examinations of their environmental, geographic, and cultural elements. Most notably, Nancy Hoalst-Pullen and Mark W. Patterson's edited work *The Geography of Beer* devotes much ink to the spatial elements of beer over a massive amount of time.[10] Although scholars have examined aspects of beer terroir, a handful have named it as such, and fewer still have attempted to combine the various elements into a single narrative.

Spatial analysis undergirds many beer histories. Due to beer's weight in bulk and the availability of transportation, a brewery's location matters a great deal in terms of distribution. Shipping by rail was the most cost-effective method, especially during the late 1800s. This was not without challenges and costs—not only weight but also the availability of rail lines. For example, in 1898, a reporter in La Junta, Colorado, decried the cost of shipping Denver beer to their city or across the country. Frustrated, they bemoaned the fact that it cost thirty cents to get Denver beer the 181 miles to La Junta while, for only seven more cents, Denver beer could go the 571 miles to Kansas City.[11]

Many beer historians have examined particular cities or regions to understand the diffusion of breweries and beer accessibility. Scholars like Brian Alberts have examined cities, while others, such as Doug Hoverson, examine statewide brewing industries.[12] In many ways, these more localized lenses allow one to explore the major events in the national beer industry. The challenges posed by proximity or distance rhyme across the

US. Connecting one's brewery to ingredients and customers—supply and demand—was and continues to be a perpetual obstacle for brewers.

Another popular approach is to follow the rise and turbulence of particular breweries. In the US, these include Coors and Anheuser-Busch among other long-lived institutions. Their longevity means their corporate histories intersect a great deal with major events in American history.[13] Beyond this, Omar Foda has traced Egyptian breweries as they navigated complex domestic and international markets in the nineteenth and twentieth centuries.[14] These venerable breweries are not alone in receiving historical attention. Much as historians today zoom in on microhistories, beer historians, biographers, and autobiographers of breweries have examined relatively recent histories of major craft breweries. Significantly, Ken Grossman and Garrett Oliver have written about the rise and reputation of Sierra Nevada and Brooklyn Breweries, respectively.[15]

Although geography and biography are useful courses, this work seeks to provide a comprehensive analysis of a region not fully contained by neat political lines. Place matters and specific, particular places have different approaches to their beer. Rather than cities, this work examines regions; rather than states, time zones. There are physical, geographic, climatic, and environmental reasons why beer brewed in certain regions is different from others. In the end, limitations and advantages vary. Water content and the use of fermentable local plant matter crucially inform the types of beers possible. Taste preference and desired outcomes also play a crucial role, as beer's flexibility allows any number of alterations, substitutions, and additions. Take, for example, the tremendous profusion of beer styles in recent years that push boundaries and expand the definition of, or reinvent, what beer means.

This work is an environmental and perceptual survey of Rocky Mountain beer. Beer, like many alcoholic beverages, is a complex agricultural and environmental product. This study does not engage whether Rocky Mountain water, local yeast strains, or homegrown barley did in fact improve the quality of Rocky Mountain brewers' products compared to their peers. Scientifically, determining how these elements would impact the beer, especially separated by so many decades, is impossible. Rather, the emphasis is on the perception of these items in how they *could* have improved the quality of the beer. Brewers, pundits, and other friends of the beer industry pitch the higher-grade quality of the local environment and agriculture as key to the superiority of Rocky Mountain beer. It is not useful, nor indeed possible, to understand if their beer was in fact better than beer made in other beer capitals, such as St. Louis, Chicago, or Milwaukee.

Certainly, a molecular and chemical analysis of the soil, barley, water, or yeast could yield a sense of the flavor profile and caliber of beer in the Rockies. Projecting these backward may garner some insight into the beer they brewed, but this is a flimsy approach to actually understanding their beer; there are countless unaccounted-for variables in this approach. Terroir is as much about perceived influences and taste as it is about realities within the glass. Trying to reverse engineer the beers historical figures consume is an interesting intellectual task. All the same, it does not address the human aspect. The concept and entity of beer mattered more in the nineteenth century than its actual quality. Here, we are trying to get at what people conceived of their beer, not whether it would have actually tasted better or worse.

Primarily focused on beer, this story will inevitably leave out quite a number of characters and stories. Beer styles will be largely restricted to lager and other beers created in Western European traditions. These are beers white Americans mainly produced as they colonized the middle section of North America. Lagers, and other Western beers such as porters, stouts, and pale ales, feature far more prominently in records from the Rocky Mountain region from the mid-nineteenth century to the early decades of the twentieth. There are a number of Native American beers and fermentables. Native peoples occupied this land far before white settler-colonists spread over the crags and valleys of the Rockies in search of mineral wealth. Arapahoes, Apaches, Cheyennes, Utes, and other groups called these mountains and surrounding regions home. They still do, and they continue their traditions and communities interspersed throughout these states. Their exclusion from this narrative does not diminish their profound impact on these landscapes.

There were, and remain, innumerable Native brewing traditions and beverages. They fermented ingredients from their immediate environment and adapted to changes much in the same way later settler-colonists would. North and South Native Americans' fermented drinks were made primarily from corn or maize, such as atole, tizwin, tulpi, or chicha.[16] Other drinks include those made from birch bark or spruce, which, along with corn, became significant ingredients in the Rocky Mountains.

Beyond Indigenous brewing, there are other aspects of brewing and terroir not addressed in this field. First are the workers and laborers within the brewhouse. Some components of their ethnic background will be covered, but the gendered, economic, and social aspects of the workforce that boiled, brewed, and barreled in brewhouses and along city streets are left out. This is not a judgment on their contribution to brewing but rather a reflection of omission by brewers. Workers and laborers were not attributed with the same terroir-bestowing characteristics as other physical and intangible ingredients. This may stem from disdain for the working class or the fact that brewers rather market their ingredients over laborers.

Workers were integral to the success of brewing, but the focus is on the messages consumed through pints and barrels of beer. Most notably, Judith Bennett's *Ale, Beer, and Brewsters* explored the gendered nature of beer and brewing in late-medieval England.[17] While gender is an important lens for studying beer, especially in the overwhelmingly male pre-Prohibition Rocky Mountains, it is not the central focus of this work. Again, beer is the main character and how we will explore the terroir of the Rockies.

Second, brewhouses' general structure and makeup and their impact on terroir are unexamined. Some major breweries, such as Coors, have significantly altered the landscapes by diverting rivers, building ice ponds, and constructing beer factory campuses. But again, the focus is on how beer reflects terroir, not necessarily how the beer was made. Separating these two may seem like splitting hairs. Most likely, in their hearts and minds, brewers considered their vats and tools of their trade as much their bread-and-butter as grain and hops. Brewers and their advertisements drew more inspiration for marketing from ingredients and intangibles than from the sweat, hands, and equipment that labored in the service of quality beer.

Brewing

Beer is an amazingly flexible beverage that has immense variables and opportunities for innovation and manipulation. What follows is a brief synopsis of brewing beer in the Euro-American tradition. This formula structures the basic brewing process and was followed by white brewers in the Rocky Mountain West.

Fundamentally, beer is the aggregate sum of water, hops, grain, and yeast. Water composes the vast majority of beer, anywhere from 91 percent to 98 percent, and brewers since time immemorial have guarded their water quality and proximity.[18] Ensuring their water source is as pure and close to the brewery as possible is crucial. Brewers start with an amount of water at least three times the amount of beer they intend to produce.[19] They then boil this water and add in malt. Typically, brewers do not create their own malt; instead that is the purview of maltsters. To produce malt, the maltsters take grain—barley is the most common—and soak it for a few days. Then the grain is laid out and allowed to germinate and sprout. During this time, enzymes and sugars develop within the grains. In order to prevent the seeds from sprouting and consuming the sugars, maltsters kiln the grains and place them into heated spaces. This truncates the germination and, depending on how long the maltster keeps them in the kiln, produces different shades of malt. Darker malt produces darker beers, and the sugars retained within the grains provide sustenance for the yeast to produce alcohol during fermentation.[20] If you have ever entered a brewery and smelled rich, toasty notes, likely malting was in process.

This malt is milled and crushed to release the sugars within the grains and added to the boiling water. This process is termed *mashing* and produces wort. After this, if the brewer chooses (and in Northern European brewing, such as in parts of Germany, this was required), they add hops to the wort and boil again. Today, brewers will repeat this process several times to get the desired hop flavor and aromas. The brew is then strained to remove particles from the hops and malt.

The penultimate step, fermentation, can occur in an open or closed container. Today, many breweries culture their own yeast strains and add them intentionally in order to replicate the same beer every brew cycle. These strains are monitored and protected to provide a high level of consistency and replicability. If breweries opt for an open-brew method, they add yeast indirectly (as they did before brewers fully understood the role of yeast in beer).

In moving the beer to the fermenting vessel, yeast in the air falls into the beverage. It also clings to the sides of the barrel between fermentation cycles. Also, some beers—particularly Belgium farmhouse or lambics—are open fermented, and yeast continues to fall in during fermentation. Yeast greatly affects the beer's character: Different yeasts ferment at different temperatures, contribute different flavors, and overall create distinctly separate beers.

Yeast and Brewhouses

Lager was the predominant beer found in the Rockies and across the US in the nineteenth century. Montanans in 1879 consumed 28,960 gallons of lager in a single July, enough for every adult and child to consume a gallon each.[21] Music festivals, much like today, were prime arenas for mass consumption. At one in 1883 Buffalo, Nevada, festivalgoers consumed 66,030 gallons *in excess* of the previous year's inebriation.[22] Collectively, lager drinkers downed over one billion gallons a year toward the end of the nineteenth century.[23] This was in spite of the growing Prohibition movement that severely reduced alcohol consumption in states like Kansas.

Lager is a bottom-fermenting beverage that ferments at a lower temperature than ales. For many beer styles, the defining factor is their type of yeast. The broad strain of yeast used in lager is *Saccharomyces pastorianus*, but there are a considerable number of other strains, varieties, and cultures.[24] Different yeasts are often one of the most defining differences between beer styles. Notably, *Saccharomyces cerevisiae* is used for ale and is a top-fermenting yeast, and *Brettanomyces* is for open-fermentation or saison beers.[25] Yeast and workers lived and toiled together to transform wort into beer. Collectively, their labor contributed to the terroir, as both were highly local to the confines of brewhouses. Sweat, breath, consumption, and other by-products built unique living environments within brewhouses.

Yeast is simultaneously omnipresent in brewing and almost completely absent in the sources. Brewers who had their own specific strains of yeast guarded them to protect the flavor profile and consistency of their product.[26] The same is true of yeast's history; although essential for fermentation, brewers were largely ignorant of its exact purpose or even existence for centuries. Over long stretches of time, brewers perfected their craft from a few germinated grains left in the corner of a granary into a well-defined and profitable industry. Certainly, they understood the fermentation process, albeit perhaps not on a microscopic level, which is where yeast resides.

In fact, the famous Bavarian Purity Law of 1493—the Reinheitsgebot—did not mention yeast, as its role in fermentation was poorly understood at the time. An omission later corrected, it was not until the seventeenth century that Antonie van Leeuwenhoek examined yeast cells under a microscope. His work was built on by later scientists, most notably Louis Pasteur, whose scientific discoveries immensely impacted the beer world and continue to define much of the science of beer.[27]

Despite its minuscule size and undetectable nature, yeast is literally everywhere, and brewers were able to get by without actively adding it to their brews. On a basic level, it converts starches found in malt into alcohol. As living organisms, yeast required places to live and reside, often on the walls, equipment, and even brewers within the brewhouse. Therefore, the terroir of not only the region but even the specific walls and tools of individual brewhouses influenced the yeast. Yeast is relatively difficult to pinpoint in terms of usage, but it opens up a useful door to understanding the tools, equipment, and physical dimensions of brewing. The buildings and equipment used by numerous laborers to perform their tasks illustrate the nature of brewing in the Rockies as small-scale brewers struggled to compete with beer factories.

By the mid- to late nineteenth century, breweries across the world were geared toward maximizing productivity and minimizing costs. Early nineteenth-century breweries were smaller and less mechanized, but by the end of the century, beer output increased by several magnitudes. Some estimates put early 1800s commercial production at 1,300 barrels per brewery; by 1900 it was closer to 22,000, with some major beer factories producing almost 1 million.[28] Inventions such as the steam engine in 1765 featured prominently in brewhouses and, along with gravity, powered operations in beer factories.[29]

Tower brewhouses were vertical structures that employed steam power and gravity. These were massive brewhouses that occupied not only the skyline but also substantial plots in cities. They held water ready for brewing on their roofs and were therefore sturdy, thick structures with serious support. Sturdy supports were essential not only to maintain the sizable infrastructure but also as fire deterrents due to the omnipresent threat of conflagration in wooden breweries. Fires ripped through breweries and had a massive role in the ascendency or diminution of breweries within cities or even between cities. Fires such as those in 1880s Rock Island, Illinois, or the 1871 Great Chicago Fire redirected the power and production of beer-producing cities.[30]

Tower breweries were common across the US in the late nineteenth and early twentieth centuries. They were the most economical and high-producing beer factories and had considerable output. In the Rocky Mountain region, there were a number of tower breweries, such as the Becker Brewing and Malting Company's building in Ogden, Utah, during the early

1900s. This brewery, in 1908, produced fifty thousand barrels of beer, consumed 1.5 million bushels of barley (all grown in Utah, clearly "said to be the best barley grown in the world"), and towered over its ten-acre campus at five stories tall.[31]

In this process, there were many specialized and generalized roles for workers. Maltsters and brewers are two obvious roles, but there were a host of laborers, teamsters, and others who ensured a smooth operation. Yeast is a microbial organism that is crucial for brewing. In many brews, such as farmhouses and saisons, the local yeast strains that float in the air and cling to brewhouse walls are essential to the style.[32] This work, however, will not dwell significantly on the workers who produced the beer. Without the beer, there would be no beer terroir. Yet workers, brewhouses, and yeast for a variety of reasons did not enter into their advertisements or vernacular in promoting beer. It seems that those marketing beer did not consider those sweating in the brewhouses a major element of the final product's terroir. Certainly, these aspects are fertile ground for future research.

1

Terroir & Taste

What stories, lessons, information, and expectations do we learn from our drinks? What labor do bottles and labels perform? Wrapped around bottles, filled with text and images, how do their artistry and information balance each other as they inform our brain about the flavors we should expect? How do they prepare us to taste certain notes, gloss over mistakes, or interpret funks and spoilage as crucial elements? Vessels contain and communicate stories, of which the label is simply the beginning. Storage and serving containers invite us to imbibe in myths and histories, nostalgia, and craft. Terroir is the confluence of human ideas, hopes, perceptions, and history along with physical demarcations, chemicals, and natural systems.

Wine is the dominant subject of terroir, and its bottles capture essential elements of terroir. The year, region, and types of grapes are not just facts but aim to build a perception of the wine. Is it from a historic estate? Do the images invoke lush country fields, majestic grand manors, or castles? Terroir provides ample fodder for advertising today, but marketing and messaging grow out of a deep, complex history. Terroir is an incredibly nebulous, intangible term that refutes clear definitions and boundaries as it seeks to encapsulate a host of practical, physical aspects in addition to sociological, historical, and cultural components. It ranges from scientifically delineated pH levels to inscrutable, intangible flavorings. Your ability to note terroir on your tongue may enhance your drinking rituals and enjoyment. It may evoke images of a long-forgotten, nostalgic, ahistorical past. At the same time, inability to pinpoint or care about terroir does not necessarily detract from the enjoyment of foodways. Infinitely nuanced, terroir is both crucial to the production and participation of drinks and summarily dismissed by multitudes of imbibers.

Defining Terroir

Terroir is the summation of culture, history, ingredients, geography, climate, vegetation, topography, and other innumerable aspects that contribute—directly or indirectly—to the flavor of foodways. Terroir is a vast world with endless nuance and complexities. This chapter will explore the avenues of terroir, historically overwhelmingly applied to wine, and then consider how to map this onto beer. It posits that terroir is an excellent method of understanding beer historically and in the present and that by examining history, culture, geography, climate, ingredients (water, grain, hops, yeast, etc.), and nostalgia, we can come to a richer, fuller conceptualization of what definitions, meanings, and contents are poured into every pint.

Explorations of wine terroir run the gamut from soil samples and deep, scientific investigation in highly specific locales to high-level studies of entire regions. For example, a team of scientists examined nuances between neighboring vineyards in terms of soil microbial content, rainfall, distance between rows, soil bacteria, and more in Barossa Valley, Australia.[1] They studied vineyard subregions by testing the bacteria contents of the soil. One of their findings was that they "identified 18 wine traits correlated with differences in bacterial community composition and diversity, and four correlated with the abundance of specific taxa," and further they "found that soil variables are the major shapers of bacterial communities."[2] This is one of many such inquiries to find concrete data to differentiate between neighboring or distant vineyards. Contained within terroir are entire worlds, and defining the term poses significant challenges.

There is a vast network of scientists attempting to quantitatively distinguish one vineyard from another. By no means a fool's errand, there are significant measurable differences between regions that impart flavoring to the final product. The same is true of all agricultural goods. Studies—scientific, historical, informal, or otherwise—attest to our human desire to separate, delineate, and define flavors and tastes that refute easy classification.

At its core, terroir is designed to differentiate between physical locations. Every place on earth has a terroir, but the degree to which they are distinct depends a great deal on the physical landscape. Culturally, it also depends on how humans have considered certain places over time. Physical borders and boundaries are easily mentally traversed if we are used to seeing them as united. Land features help divide or unite places, and how humans historically considered space matters just as much. Indeed, much of our understanding of distinct regions stems from years—even centuries—of producers and marketers convincing us of binaries instead of unities.

The Rocky Mountain region, as a unity, is much a creation of historical memory and construction but could have just as easily been divided into demarcated, segregated spaces. Here, we will take a broad approach, as we are attempting to show how the Rocky Mountain region's beer terroir is different from, for example, the American South or New England. The difference is both scientific and social and, in many ways, constrained by perception.

What one includes in a description of terroir greatly depends on the case they are trying to make. To provide a summary of the various approaches to wine terroir is beyond this work, as it is an immense subject that consistently fills volumes. Rather, the intention is to examine different conceptions and frameworks used to approach the world of terroir and map useful approaches to beer terroir.

Scholars have expended considerable time and trees considering terroir's limits and applications. Peter Howland and Jacqueline Dutton, in an edited work that examined terroir as a function and lens to understand utopia, consider whether terroir is the cumulation of the conditions the wine was produced within. In this, they include "site of production, seasonality and varied *vini/viti*-cultural interventions," and further, the "specifications and variations in the bio-historic constructions of taste."[3] In the same work, Robert Swinburn's case study of Southeastern Australia aimed to expand terroir beyond "the complex intersections between political dominance, social class, and economic monopolisation [sic]."[4] In addition to these boundaries, Swinburn considers terroir a form of utopia—that is, a concept that includes "minor realities" alongside a higher ideal that lives within these realities.[5]

Other studies of wine and terroir link it more explicitly to identity. This connection, explored by Matt Harvey, Leanne White, and Warrick Frost in their edited volume *Wine and Identity*, provides terroir with a personal dimension.[6] In this case, terroir applies not just to the land, or even its products, but to the people in the vicinity who cultivate the land. People's literal blood, sweat, and tears enter the soil and connect them to the soul of the land. At some point, however, it is useful to stop and ask, To what extent does terroir end? At which point does one vacate one terroir and enter another; what categories of analysis are excluded from terroir? There are as many cases for continuity as for contraction, and scholars have expanded the avenue through which they explore terroir.

One examination, led by Percy Dougherty, uses a handful of lenses and case studies to explore terroir. Their work, *The Geography of Wine*, parses out wine terroir into several categories: regional, physical, cultural, economic, and techniques.[7] In this work, a collection of scholars discusses the values of using terroir, issues in defining it, and how case studies challenge or confirm delimitations of terroir. They are not alone in attempting to establish parameters to partition off terroir in an effort to describe and encapsulate the nebulous term. Their analysis is useful, however, in the emphasis they place on geography, a category that itself contains wide and deep implications.

These are just some of the ways to understand terroir from the top down. From the ground up, we can examine the layered meanings of terroir. Fundamentally and linguistically terroir is soil.[8] Terroir includes the physical materials in the soil, the historical and cultural influences that dictate expectations, the actual production and flavor goals, and the memory infused into the product.

Contained within the soil are chemical, geological, climatic, and human-influenced components. Soil is not a neutral state but rather the product of millennia of natural and human forces. At the onset, terroir is a physical attribute produced by intentional and unintentional climatic and cultural events. In terms of wine, vineyards worked generation after generation have had their soils maintained or manipulated in overlapping and complex ways, and although the geographic boundaries may be the same, the soil itself evolves over time.[9] Grapes and grain extract nutrients and moisture from the soil as the plants grow. Agriculturalists counterbalance this as they inject nutrients back into the soil for continued growth.

Supported by foundational soil, historical and current human values and dreams manifest. Land and space are canvases upon which human drama occurs. Drawn from physical, social, and cultural landscapes, values and ideas about how a certain product should taste inform terroir. In wine, products from certain regions are supposed to taste a certain way, transform from grape to wine in a prescribed manner, and evoke certain historical characteristics. Likewise, beer styles invoke methods, tastes, and histories of production that harken back either to true lived experiences or to a manufactured halcyon era. Taste is as much a product of history and culture as soil and rain.

On a higher level, terroir also inspires and evokes worlds of potentiality. It is not only a measure of realities on the ground but also an ideal of what might be. Terroir is an aspirational target that producers aim to reach; it is a series of characteristics and qualities that producers desire to fulfill. This goal is never fully realized. In each batch created, producers consider how close they came to their perfect, and unattainable, goal. Each batch is the embodiment of their best efforts to reflect their current terroir. It signals the status of that particular terroir and begins the process anew by fostering a new ideal terroir to reach in the next iteration. In this way, terroir also incorporates craft and artisanal elements defined by the cultural goals for the product. In terms of beer, the brew desired is not just palatable but one that conforms to an expected taste, flavor, and production profile.

Geography and landscapes connect to our vision of what a place once was. Memory and nostalgia are aspects of historical connection to a place. History is a battleground for ideas and something to which people have continually turned to forge identities. Terroir-informed foodways are no exception to this practice. One notable late-nineteenth-century French example comes from the region of Champagne. After years of agitating, litigating, and building a collective identity, the region won the sole rights to use the moniker *champagne* on its titular wine products.[10] To formulate the brand and to make the case for their sole ownership of it, the trade body Grande Marque made historical arguments as to the authenticity of champagne production in that hyperspecific locale.

Beer Terroir

These debates, perceptions, and realities can help us understand any food product. Here, this approach complicates our understanding of beer and brewing. This is not the first analysis to examine the terroir of beer, though perhaps one of the first to examine beer terroir as a totality rather than in terms of single elements. It is a rich and wide topic that warrants several approaches and avenues of exploration. At its most basic, beer terroir continues the tradition of connecting a place to a product.[11] In the same way as wine and other foodways, beer terroir needs to convey authenticity and evoke—through branding, imagery, and narratives—why that beer specifically benefits from the exact location of its production.[12]

Beer complicates terroir in ways not represented in many other alcohols. The ingredient list often draws vegetable matter from various fields that could be quite distant. Hops could come from several varieties and countries; grain from various fields; additives from near or far regions. In this way, beer terroir represents a blending and combination not seen in wine or spirits, even in their own blended versions.

Borrowing concepts from wine terroir, beer terroir also contains multitudes. In this narrative, we will define beer terroir as the following: It is the cumulation of overlapping multitudinous historical, climatical, geographical, cultural, and flavor contributions to beer. These factors are regionally specific and can become intensely localized with additives in terms of ingredients and meanings. Further, beer terroir encapsulates entire regions, as agricultural and cultural ingredients are imported from great distances. In the Rocky Mountain region, beer terroir incorporates the advantages or disadvantages the region bestows on brewers and their craft, what must be imported, and the historical forces that facilitate a specific type of brewing.

Many previous beer and beer terroir studies have become enmeshed with hops and their terroir. A great deal of studies have considered the scientific aspects of hop terroir rather than the social or historical components—meaning the focus was on what was scientifically and demonstrably different about types of hops rather than on what set them apart perceptually or historically. This is not to diminish scientific reasons for why certain landscapes produce a discreet product. Again, fundamentally, terroir is about soil from which plants grow, and thus even slight microbial differences can lead to distinct flavor divergences.[13]

Hops provide an easy entrance into the world of terroir for beer, and it is understandable why they have attracted so much attention. Because of the qualities they impart to the finished beer, brewers today are highly cognizant of where their hops originate. This is as true now as it has been historically. Modern brewers' abundant choices of hops, or other ingredients, should not diminish the similar calculus nineteenth-century brewers performed to decide which to buy and use.

In the Rocky Mountain region, brewers balanced hop quality and cost. Preferred hop-growing states varied among brewers. For example, along the late nineteenth-century Colorado Front Range, brewers debated the relative merits of hops from the West or East Coast, particularly California or New York.[14] Brewers had to weigh the price of ingredients against their beliefs in the superiority of one source versus another.[15]

It is not sufficient to examine only why or how certain locations impart specific flavor profiles or support certain strains of hops. To gain a greater sense, we must understand how and why hops were grown in the first place. In the Rocky Mountains, hop species do not grow well and were never produced in great abundance.[16] However, brewers imported hop cones packed into bales in great numbers every year for much of the last 150 years to support a brewing industry not endemic to the region.[17] Human colonial settlement has changed agricultural and brewing patterns, and the white settlers who came to the Rockies transformed the terroir to incorporate beer. The point here is not to investigate hop terroir—that will be discussed further on. Rather, there are many ways that beer, with its multiple steps and historic provenance, benefits from a terroir-based framework.

Beer terroir is the summation of a million variables. This book seeks to address a handful in a particular region. There are innumerable aspects left out, and inclusion in this work attempts to address some of the larger, more influential components. Any number of factors influence the end product of beer, and here the goal is to cast a net wide enough to capture the ones that impart the most to the finished beverage. At the same time, this must be balanced against the historical record. Brewers did not mention, or were not aware of, everything that went into their brews.

More geographically, there are countless subregions within the Rockies that will have a separate terroir, distinct from the greater whole. However, the mission here is to apply terroir broadly to brewing and to show how people conceived of, and created, terroir in their region without naming it as such. In this way, terroir is an integral, unseen, yet common component of brewing.

Brewers, pundits, boosters, drinkers, and any number of people involved remotely with beer worked together to build beer terroir in the Rocky Mountain West. They did not necessarily intend to collaborate, but their efforts to push their local beer to the forefront brought them together with others in the region with the same goal. Their language, goals, history, methods, and physical and climatic environments mirrored their counter-parts across the region. It is through the way in which people in the Rockies viewed their beer that we see a congruent terroir evolve.

It is important to be aware that many of these efforts were largely unintentional. That is, while brewers promoted their beers and presented their businesses in a certain way, there was not a concerted effort to build a Rocky Mountain beer terroir. Certainly, some brewers were more visible in building a narrative of terroir, and brewers borrowed useful tactics from one another. In this way, brewers realized that using aspects of terroir could sustain their business. Piece by piece, then, Rocky Mountain brewers con-structed a framework of terroir particular to their place, though they would not have labeled it as such.

The effort here is to use terroir to highlight beer history and to center beer. This approach differs from other beer terroir efforts that zero in on specific ingredients—most commonly hops. Although the aspects of ter-roir covered in this work are separated by chapters, it is crucial to remem-ber they work in conjunction both to create beers and also to construct a regionally specific terroir. Each beer contains terroir and a summation of human choices and ecological constraints.

Terroir is the lens through which we will examine brewing patterns in the Rocky Mountain region from the mid-1800s until Prohibition in 1921. The ensuing chapters tackle separate aspects of terroir. The first part of this book addresses physical dimensions, such as climate and geography, as well as provides historical context as to the types of beers under discus-sion. Since beer is a product of multiple ingredients, the second batch of chapters is dedicated to the main ingredients in beer—water, grain, hops, and yeast. Finally, history and nostalgia contribute to and perpetuate ter-roir's power, and the penultimate chapter delves into the uses of memory in advertising in the last few decades.

2

Climate & Geography

Drinking is an agricultural and geographical ritual, consuming liquid sunlight and earth. This is true no matter the substance. Energy provided by the sun and nutrition supplied by the earth coalesce in innumerable ways through natural and human-driven processes into a state of matter that satiates and quenches. In this way, when we consume liquids, we are tasting the provenances of the collective ingredients. Whether they are factory-fabricated minerals or hand-pressed grapes, they all come with the terroir of their place of origin.

In wine, there are any number of climatic elements to terroir: the direction in which the vines are arranged, humidity, sunlight, soil type, pH levels, rain, and so forth. Some of these are within the realm of viticulturists to plan and control. Others are firmly outside the power of humans to manipulate. The give and take between human and environmental forces establish the battleground on which terroir plays out. The environment equally aids and thwarts efforts to describe and define terroir. It presents unique opportunities to build upon while imposing real, strict, and occasionally insurmountable barriers.

The same categories apply to beer, but with a slightly different focus. The impact of the climate on water, hops, and grains will be discussed in other chapters. Here, the focus is on the landscapes and environment upon which brewers built their enterprises. These are natural facts that, despite heavy environmental manipulation and degradation in the West, ultimately do not change. The climate and geography posed challenges that brewers occasionally turned into advantages. If they were not real advantages, then brewers underwent a series of public relations efforts to posit them as such.

Yet before they could do that, brewers trekked a difficult path that more often than not led to failure. Understanding their new home, its limitations, and its characteristics is crucial to brewing beer and to understanding the foundations of terroir.

The Rocky Mountains are a specific region within the American West, one that can be divided exponentially into infinitesimally numerous ridges, peaks, parks, valleys, and trails. It is a space that has defined American ideals and dreams while itself containing a litany of transgressions and violence.[1] At once peaceful, romantic, and majestic, within these mountains, untold blood has been shed in the name of gold, land, and white supremacy. Defining the American West as a cultural, historical, geographic, or otherwise contiguous unit is quite difficult. Most notably, historian Patricia Limerick has detailed the West as an arid, vast region that has been home to a diverse range of people and to boom/bust industries and is "particularly prone to demonstrate the unsettled aspects of conquest."[2]

Mountains as discrete, definable entities pose additional conceptual problems.[3] What makes a mountain or a range? At what point does a series of peaks make a range, or at what point do the mountains merge into some other landform? Beyond the physical demarcations of a mountain range, what are the social or cultural limitations? Even if people do not live in a strictly literal sense within the Rockies, at what point are they oriented toward the peaks and incorporating that within their community and identity? Just as in the case of terroir, conceptualizing a region as a united whole is rooted in its cultural, historical, and natural characteristics as much as it is by people's—historical or present—perception of that region as a discreet place.

These are questions that we will consider throughout this chapter and beyond. Geography and environmental boundaries are the clearest form of terroir. Wrapping our minds around physical space is the first step to considering the factors of terroir within that space. This chapter will outline roughly the geographic confines of the Rocky Mountain West. In doing so, the aim is that slight differences and regional variations will come into better light. The physical characteristics of each region share some similarities, but brewers faced climatic differences and varying access to environmental resources depending on which region they inhabited—problems experienced by individuals who in turn found kindred souls across the Rockies. Despite these fluctuations in rainfall or elevation, in the end, this region is a coherent whole that beer helpfully knits together.

Mountain ranges form, rise, fall, and regrow over unimaginably long periods of time. The Rockies as we know them today are the result of serial mountain productions, destructions, and regenerations. Earth's tectonic plates carry the crust and the mantle, together comprising the lithosphere. The crust is the surface level of the earth, with the mantle below. These gigantic floating islands crash, collide, and crumble into one another as they float upon Earth's core, driven by extreme heat.[4] When plates smash into one another, some sections are lifted while others are subsumed underneath. This mountain-building process takes millions of years to form peaks, which in turn take millions of years to erode down. The Rockies were built over several waves of mountain building. Most recently, the landforms that Americans call the Rocky Mountains formed during a seismic event called the Laramide Orogeny.[5] Occurring around sixty-five million years ago, these mountain formations grew incredibly slowly into the range they are today.

The result of these titanic collisions and the ensuing upsurge of earth and rock created a range of mountains stretching from the northern tip of Canada to the southern edges of the US. Climatically, culturally, and historically, the range varies in its latitudes. Here, we are concerned with the American Rockies rather than their Canadian counterpart. A few reasons for this: Simply, there must be delineations to terroir, for fear it will become an unwieldy unit too large to make worth quantifying. Every place on earth has its own distinct terroir. How we quantify and organize this is an argument about terroir and an evaluation of shared or differing characteristics.

As such, although environmental regions do not nicely adhere to political boundaries, in this case, we will truncate the range on the American side of the Canadian border. American brewers in the nineteenth and early twentieth centuries aligned themselves toward, or in competition with, American cities. In part, their shared identity was built in their relation to other American cities. They were focused on American markets and consumers, particularly after the 1893 McKinley Bill, which drove up tariffs and effectively severed Canadian barley imports.

Separating Canada from the US solves the northern geographical problem but does not address the east-west barriers of Rocky Mountain beer terroir. Defining the American West has occupied historians and scholars for generations. Simultaneously, there are clear demarcations—for example, along the Front Range where the Great Plains meet the mountains—while at the same time, there are no clear boundaries to these peaks. With such a vast region with infinite capacity for subdivision, this is inevitable. Yet I hope this study spurs further research into how brewers operated within smaller pockets of the Rocky Mountain West.

The physical dimensions of the Rockies are the foundation upon which terroir grows. The shape of the mountains affects air current and precipitation patterns, and the process of mountain building fundamentally changed the soil and available plant and animal life. In so many ways, brewers who recently relocated to the Rocky Mountains found their environment at odds with how they were used to brewing. Aridity and altitude, quite different from the American East Coast, England, or northern Germany, meant that brewing and preserving beer posed challenges they had not come up against before. While the dryness meant that beer would not be spoiled quite as easily from humidity, the altitude decreased the temperature at which water boiled, requiring greater vigilance, though less fuel.

Map 1 overlays the geographic upon the political. For the purposes here, the states within the Rocky Mountain region west to east are Nevada, Idaho, Utah, Montana, Wyoming, Colorado, and New Mexico. This captures the heart of the region, particularly pre-Prohibition. These states were among the more populous during this period and in many ways have captured the concept of the American West. Further, we can use this list of states to compare and contrast beer terroir.

States included in map 1 but not under discussion here are Arizona, California, Oregon, and Washington. The reasons for exclusion vary by state. Arizona and California are excluded because the regions incorporated in this map were extremely sparsely populated at this time. For example, Arizona between 1870 and 1900 grew from slightly under 10,000 people to around 122,000. During the same time, its neighbor Nevada went from 42,000 people, peaked in 1880 at 65,000, and then dropped back to 42,000—this time a few hundred below their 1870 totals.[6] Population is not the only reason; also significant is what we consider the Rocky Mountain cultural world. Arizona's beer and brewing industry was minuscule and did not participate in the same brewing world that other states included in this study did. The states included were in conversation and competition with a similar group of breweries and drinkers.

Oregon and Washington certainly know their beer and had their own conception of terroir. The famous Olympia Brewery, based in Washington, was massively influential and only recently closed after 125 years in

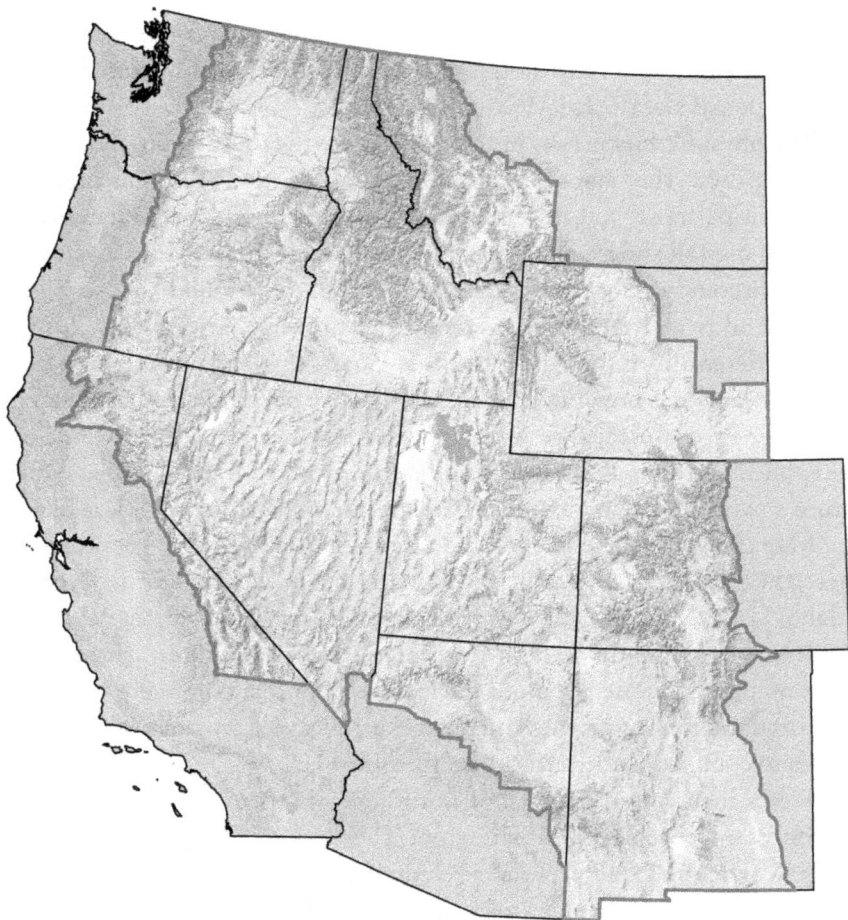

Map 1: Northern Rocky Mountain Science Center, "Map of the Intermountain West." Public Domain. https://www.usgs.gov/media/images/map-intermountain-west.

business.[7] They advertised into Idaho and clearly understood the terroir of their own region. One of their advertising slogans promoted, "The water makes the beer," and said that the beer was "made from clear sparkling spring water."[8] However, the Pacific Northwest has a distinct regional and cultural terroir distinct from the Rockies. Although they are along the western side of the American and Canadian Rockies, they face west toward the Pacific rather than east toward these peaks. Their climate, culture, and other factors lend themselves well to a separate examination of terroir.

Map 2, "Physical Regions of the Rocky Mountain West," perhaps lays out a better physical understanding of this region and participating states. By showing the climatic regions, this illustrates how they are in line with concepts of terroir. As we see in this map, the spine of the Rockies and the core of the region include parts of northern New Mexico, the majority of Colorado, Utah, and Wyoming, and sections of Idaho and Montana. These areas are the forested mountain ranges where prospectors went to find their riches

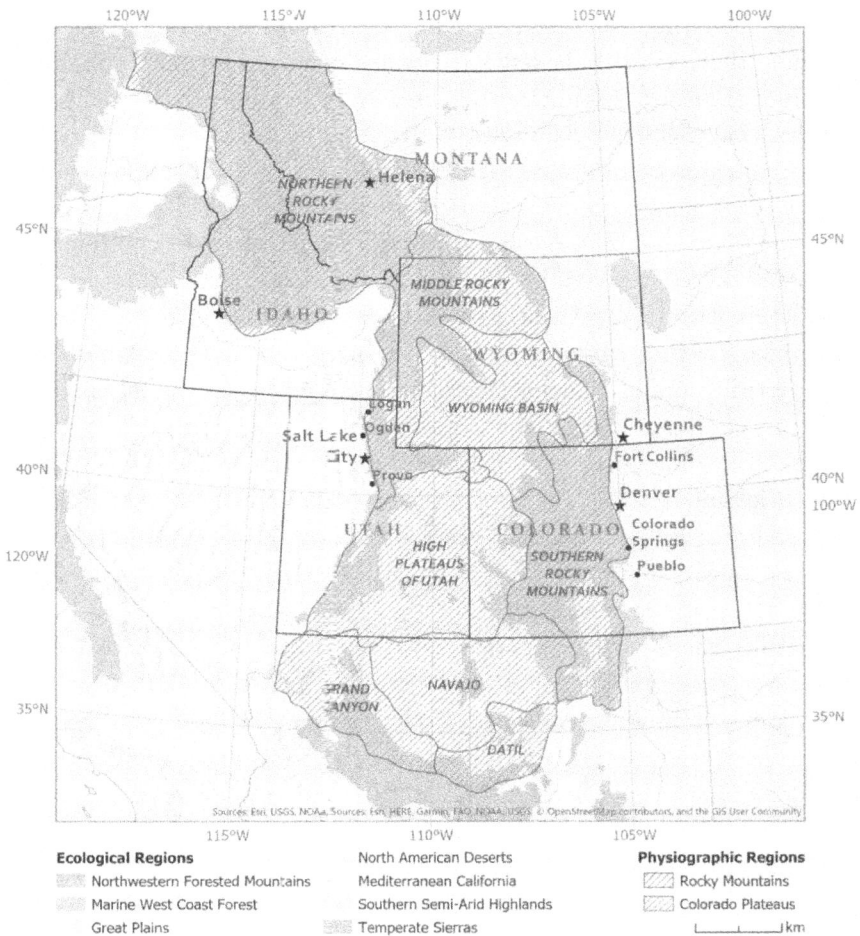

Map 2: "Physical Regions of Rocky Mountain West." From *The Rocky Mountain West: A Compendium of Geographic Perspectives*, preface, by Michael J. Keables (2020), copyright American Association of Geographers (http://www.aag.org).

and brewers followed to find theirs. Similar to mines that boomed and busted in quick succession, breweries in the mountains or even in cities like Denver overwhelmingly did not celebrate their first anniversary. Proudly in the directories one year, silently absent in subsequent years.

In some ways, the states under consideration handled problems quite differently. Yet the challenges they face—here in terms of shared climate and geography—show us how, regionally, they shared a sense of beer terroir. Also, the point here is not to define strict geographical boundaries based on terrains but rather to understand and weave together threads from across a region. Brewers in these states individually and collectively built a beer terroir of the Rocky Mountains by plugging their particular beers into this system. They highlighted advantages, shared by brewers across the Rockies, that their region bestowed on their beers and bemoaned shared difficulties.

It is important to note throughout that although brewers did not always see themselves as a collective unit, their shared characteristics both in their beers and ingredients and in their mindsets indicate a special, shared terroir. There certainly was not a singular conception of terroir posited by the brewers of this region. Instead, we see brewers employ the same tactics in advertising and brewing to promote their beer. Of course, all breweries today and in the past endeavor to position themselves as producers of the finest beer through their selection of the most wholesome ingredients. Yet the crucial difference here is the desire to promote one's product in tandem with real geographical, historical, and agricultural constraints. These hindrances brewers in the pre-Prohibition Rocky Mountain West shared and sought to overcome in similar ways.

Getting a sense of precipitation in the late nineteenth and early twentieth centuries is difficult. Ridges and mountain spines, depending on their orientation, alternatively break or accelerate rain and wind patterns. Certainly, there are records that go back that far, and we can use aggregates over states and long periods of time to smooth over anomalies. The Western Regional Climate Center (WRCC) maintains records from the 1930s and aggregates them over thirty-year periods. The ensuing numbers come from their 1931–60 averages. By no means perfect for capturing the pre-Prohibition period, they do capture a sense of the aridity and precipitation through the Rocky Mountain regions.

The Great Basin is a massive region that drains the western side of the Continental Divide. This is the westernmost subregion under consideration here and includes the majority of Nevada and sections of Utah. Nevada's Great Basin National Park is within this tremendous region and hugs the eastern border with Utah. The average rainfall for this park is extremely low. March is typically the rainiest month with 1.4 inches of rain; several months tie for driest with 0.9 inches.[9] They can receive considerable snowfall, the maximum reaching forty to fifty inches in the snowiest years. However, this desert region is dry with low humidity. The snowpack feeds many of the region's rivers, which drain away. Inhabitants achieved water retention by laboriously diverting and trapping water sources.

The Western Slope is just that: the slope of the mountains that faces west. It runs along the western edge of the Rockies, where they run against the Great Basin. It includes the spine of Idaho, Utah, and the edge of northern New Mexico. Colorado itself has a subregion called the Western Slope, but that is a state demarcation rather than a transregional one. In map 2, we see the continuation of climatic and geographic regions between western Colorado and eastern Utah. This region receives far more precipitation than the Great Basin, though it's still a quite arid place. Nevada, which overlaps a great deal with the Great Basin, only receives a few inches a year.

Using the WRCC dataset, over this period, Nevada likely received about 8.25 inches per year. In contrast, the Western Slope states—New Mexico, Utah, Idaho—received 13 inches, 11 inches, and 18.5 inches, respectively.[10] The moisture in these states comes from the eastward-bound winds pushing up against the mountains. After collecting water over the Pacific and collecting what was available in the Great Basin, the clouds are heavy with water. Unable to rise over the mountains with such weight, they are forced to drop water as rain and snow.

The third section of the mountainous region is the High Country. Here are the mountainous regions of the Rockies, the ones from which the mountain range earns its name. Sections of Colorado, Wyoming, and Montana, along with the eastern side of Utah and Idaho, compose this section. This is, in many ways, what people think of when they consider the Rockies. Within these historic rocky outcroppings, Utes, Arapahoes, and Cheyennes traversed in pursuit of food and water; during the period under study here, prospectors and miners crawled through these slopes hunting for mineral wealth. This region is the highest in altitude, receives the majority of precipitation as it blocks moisture moving west to east, and is where numerous isolated mining camps boomed and busted.

Finally, crossing west to east, we reach the Front Range. Of the regions covered here, this was the most populated with the most stable population and production centers. Communities located at the border of plains and mountains, like Denver, Pueblo, Santa Fe, Cheyenne, and others, acted as crucial waypoints between geographies. Laborers, supplies, food, and beer traveled west into the higher altitudes, and gold, lead, timber, and other raw materials came down in elevation. Over time, these towns competed fiercely for superiority along the Front Range. They sought to control the supply into and out of the mountains as well as to diversify their town to make it resistant to the horrific cycles of the mining industry.

The Rocky Mountain West contains some of the highest-elevation communities in the United States. Major cities such as Salt Lake City, Utah, and Denver, Colorado, sit at roughly four thousand and five thousand feet in elevation, respectively. These towns were major depots through which raw materials and goods passed back and forth to higher elevations where much of the mining and resource extraction, and thus brewing, took place. Mining towns like Taos, New Mexico, and Leadville, Colorado, rest comfortably at slightly over nine thousand and ten thousand feet above sea level. Innumerable scattered communities and encampments sprung up at the highest reaches of the Rockies as people trawled in search of mineral wealth.

It is within this massive region that brewers built terroir as they labored in their brewhouses. Unintentionally, through their daily battle for profit, they developed a cohesive, united terroir based on their climatic and geographical locations. Situating themselves in their new homes was not an easy task, and they faced considerable demographic, environmental, economic, and other challenges. However, over time, certain towns grew more rooted—and the breweries along with them. Piece by piece, as white settlers colonized and transformed the human and natural landscape, brewers created and transported their craft and terroir.

3

Pure Mountain Water

Conceptions of wine terroir regarding water involve rain, aridity, humidity, groundwater, and the various manifestations of water and moisture. Further, they consider these elements in how they impact the grapes' growth and consistency. For beer, we can take these considerations and amplify them in new ways. Water not only matters in terms of growing, which we will cover in chapters on hops, grain, and other ingredients, but also is by far the largest component of beer. Wine—unless it is diluted—does not contain any water additives and merely comes from the moisture of grapes. Beer, by contrast, is made by fermenting ingredients in water, which then becomes the beverage. Upward of 90 percent of beer is water, which does not account for the water that does not make it into the finished barrel.[1] In this way, water provides several layers of analysis for beer terroir; here we shall be interested in water availability and quality for brewers.

Water, precious and central to life, is a heavily problematized commodity in the West. Physical attributes of the region hinder ready access to water, and power and survival often came through control of major streams or rivers.[2] Compounding this, industries such as farming, mining, and ranching have drastically altered the water tables, natural flows, and content of the Rocky Mountain water region. As these land uses became more industrial, their commensurate impact skyrocketed. The Rockies provide a fertile place to understand how regional differences in climate amount to significant disparities in water access, quality, and sources.

In the US, wind patterns push moisture west to east. The Colorado Basin is incredibly dry, but once clouds run up against the Western Slope of the Rockies, they begin to discharge much of their water contents. Depending on the time of year, this can be either rain or snow. As wind pushes clouds against the Rockies, there is increasing pressure that pushes clouds higher in elevation. In order to climb in elevation, they discharge moisture. Clouds continue to reduce weight by releasing water, and by the time they pass the mountainous terrain, they have lost a considerable amount of their initial water content. The Front Range, on the east-facing side of the Rockies, receives little water compared to the western side, as clouds have lost much of their moisture by the time they surmount the mountain peaks.

Water availability and quality translated into difficulties for brewers. Yet these became central to terroir as they found ways to promote their beer in spite of these problems. There are two major components when considering water in beer terroir. First, water in the Rockies was scarce and sporadic, alternating between periods of drought and overabundance. Even on the Western Slope, where clouds deposit moisture in order to rise in elevation, there is not an overabundance of water. As such, acquiring, transporting, and diverting water was a major occupation of people living in this region. Second, this water was of dubious quality, as ranching, mining, farming, and natural mineral content brought with them much contamination and pollution. These two groupings, coupled with a growing temperance movement in the late nineteenth and early twentieth centuries, made it rough going for brewers who wanted to promote their beers. In the end, brewers drew heavily from the well of pure mountain water advertising to promote their beer and ultimately established a baseline for water in considering beer terroir.

Water Availability

One of the enduring issues in the history of the western US is water. Tomes have been written tracing the various issues of control, need, use, and ensuing conflict—as often political and legal as physical and deadly. At the high altitudes of the Rocky Mountains, although there is considerable precipitation through winter snows and summer rains, much of it flows quickly downhill away from many mining camps or towns.[3] In this climate and region, precipitation did not necessarily mean a sustainable water supply.

The nature of the water cycle meant there was a disconnect between brewers and the nearest water supply. From grain to gallon, beer production today takes a tremendous amount of water—somewhere between eight and twenty-four gallons to produce a pint of beer.[4] Much of this water use is to grow the grain and hops, but between mashing, wort, lautering, and other steps, there is significant water dependency. First and foremost, water supplied the vast majority of beer's substance. Second, for beer factories, it provided water for steam engines to run. Third, proximity to water, especially in these higher elevations, allowed brewers to cut and store ice needed for storing lager over the summer.

This water usage made it critical to have access to water systems. Water's weight makes transportation an expensive and complicated issue. Typically, humans have settled close to water sources and thus negate much of the distance between brewer and water.[5] In the case of the Rocky Mountains, however, the settlement of the region by white settler-colonists during rushes for mineral wealth meant one particular resource prescribed where people lived. Gold, silver, and other precious metal veins did not always align with readily accessible water. Prospectors chose their locations not based on water, as had historically been the case. Thus, they—along with hopeful town planners—had to bridge the spatial gap between settlement and water.

Crisscrossing this region were a number of rivers, creeks, and other water systems. Much of the water fueling these streams came from snowmelt. Seasonal streams provided many opportunities to collect snowmelt, and larger rivers, such as the Colorado, dependably provided water year-round. One consistent need of the larger breweries, especially along the seasonally warm Great Basin and Front Range, was ice during the summer that allowed for year-long lagering and brewing. Breweries with established operations and clientele routinely redirected rivers and streams to form small ponds, from which ice was cut and stored in immense icehouses to keep for summer brewing.[6]

There is no end to debate and conflict over water access in the West; here we are concerned with how brewers acquired water and how they fit into this struggle. Of the many issues, two are particularly important for beer terroir: scarcity and quality. Water comes to the Rockies through a two-part annual cycle. In the winter, snow accumulation contributes to snowpack that, as it melts, provides considerable water during warmer months. This water varies in flow; it can trickle at times or flood and wash out whole valleys. During the summer, a weather pattern called the North American monsoon distributes rain from the Gulf of Mexico and Pacific through the region.[7]

In addition to seasonal variations, there are longer-term fluctuations. El Niños create increased rainfall and often flooding. These patterns are born from wind blowing across the warm Pacific waters and last between six to eighteen months. Conversely, La Niña has reduced precipitation due to winds coming from cooler sections of the Pacific.[8] In all, water sources in much of the Rocky Mountain West depend on snowmelt to refill aquifers and fill streams as well as on summer rain to carry them until the snows return. A few breweries worked around this by building near active springs. Wagener's Brewing Co. produced their Imperial Beer with "our own crystal spring water–100 per cent pure."[9] This beer was proudly "Brewed in the Mountains" with its accompanying beer terroir.[10]

The variability of water throughout the year, its inconsistency, and widely distributed sources mean water reliability is never determined. Channeling and controlling this water was a key mission of nineteenth-century white settlers. In the early days of settlement and colonization, acquiring water was an essential task. Prospectors established camps near mining sites, which grew into towns and then cities as the mines continued to produce ore and the industries within the city diversified.[11]

Fluctuating precipitation meant irregularity in water supply. Not enough water followed by a deluge caused many problems. Weather forecasters in Carson City, Nevada, worried in spring 1898 about the small amount of water trickling down from the mountains. They further worried about late spring, early summer snowmelt coming down in such a rush that farmers would be unable to divert enough for irrigation, and instead the water would rush past their dry fields.[12] Simultaneously, they found water inadequate and inundating.

Farmers' plight affected brewers' supply of grain, and their needs combated those of city industries. For brewers, however, they were also concerned with water access in population centers. More central here, however, was water for cities and towns. Commercial brewers were more likely to be in towns with access to resources and customers, and therefore municipal water supply was crucial to their success. Diverting water and otherwise connecting people in towns to water was a constant labor, and systematic municipal water supplies required years of work and logistical planning.[13]

There were two main paths to acquire water: dig wells or divert water flows to reservoirs. These were continual efforts to keep water supply up with burgeoning city demand. City boosters hoped to grow their newly created hometowns and eagerly promoted an expanded water supply to maintain populations. Unclean or inaccessible water was certainly not the only reason towns floundered or disappeared, but it was a significant deterrent to growth.

With sufficient groundwater, wells in one's municipal backyard could provide relatively quick, easy access to water. However, groundwater in many places in the West is far below the surface and refills slowly. Tapping into the groundwater could mean several hundred feet of drilling, and there was no guarantee of success even in neighboring wells.[14] In 1905, a Mr. Brown in Salt Lake City, for example, drilled a well four hundred feet below his property and hit a water source so profuse it provided water to his house and business and still ran a large stream with excess water.[15] Bonanzas of water could be more valuable than their gold counterpart.

Yet this success story was not without many caveats. Supplying water to a city was a multipronged, full-scale effort. First, digging wells did not always result in such fine geysers. Further, groundwater supplies relied on adequate snowfall from previous winters. Even then, the recharge rates throughout much of the West are minuscule. For example, one study on the Front Range estimated that three-quarters of precipitation was lost through evaporation and only approximately two inches per annum entered groundwater deposits.[16] Second, to supply a city with enough water, single wells would not suffice. The pundit writing about Mr. Brown's success laid out a plan for the city that involved wells, a system of pumps and canals, and a desire to ensure enough water year-round to avoid a "water famine" that the city had suffered the year before.[17] A secure supply relied on luck, labor, and constant restructuring.

Ultimately, the plan was intended to keep the water within Salt Lake City confined as much as possible. Beyond the need for water for the city dwellers' purposes, control and ownership of water were as crucial as its cleanliness. At one time, a Salt Lake City newspaper articulated just this, saying, "It is pure water, it is right here within the city, it belongs to the city"—a clear statement to the jurisdiction and property rights over water.[18] Statements like this could be read, heard, and legislated across the Rocky Mountain West during this era and even to the modern day. Water ownership in the West was an extremely contentious issue and was the perpetual source of conflict between rural and urban enclaves. To provide a settlement with water was not only a physically demanding task but also a financially and politically costly venture.

The above were issues of quantity. Brewers were much at the mercy of their municipal leaders and their own coffers or backs to acquire water. Across the Rocky Mountain West, aridity plagued brewers desperate for sufficient water. But it was not simply about acquiring water. The purity or pollution of the water received mattered a great deal. Brewers faced the quality of that water from its process through soil, dirt, rock, and the microscopic materials it acquired during its journey. Filtering through the soil is typically a purifying process, but as ranching, agriculture, and mining ramped up, ensuing waste contributed unwanted and unhealthy particles.

Drys & Wets

The American temperance movement has long and deep roots. Famously, Dr. Benjamin Rush issued a diatribe against alcohol in 1784. In his work *An Enquiry into the Effects of Spirituous Liquors upon the Human Body*, Rush laid out several points against the consumption of alcohol, at some points comparing the ravages of war as less harmful to the body than the consumption of alcohol.[19] Prohibition in America challenged all brewers and publicans, and in the Rocky Mountain West, it formed a common enemy for breweries to manipulate. Situating themselves against prohibition, but also not as harmful as hard liquor, brewers and beer-boosters walked a thin line in the fight against temperance and prohibition.

In the West, prohibition took shape and gained traction on a municipal and state basis far earlier than other regions in the US. That is not to say other states did not enact prohibition earlier—such as the infamous Maine Laws or Kansas mandating statewide prohibition in 1881—but as a region, the Rocky Mountain West had a particular problem with prohibition.

Although the Eighteenth Amendment was not passed until 1919, by the late eighteenth century, prohibition was making inroads in the American Rockies. In the 1894 Colorado general election, the Prohibition Party offered candidates for the supreme court of Colorado, governor, lieutenant governor, secretary of state, treasurer, state auditor, attorney general, superintendent of public instruction, two for regent of the state university, and for a district judgeship.[20] They were not alone with broad Prohibition candidates. A few years earlier in Montana, half of all electable positions had a Prohibition Party candidate.[21] This pattern reflected statewide and local elections across the region.

These political hopefuls represented a growing movement in the Rocky Mountain West. Heavy drinking and ensuing violence, poverty, health issues, and other social and moral concerns were rife in the region—at least in the news and on paper. Against this fear, drys sought to curb access to alcohol in order to reduce drinking. A leading voice across the nation, the Women's Christian Temperance Union (WCTU) had chapters across the Rockies, such as the WCTU of Montana, with eight local chapters and the power and funds to print statements of their goals, the ways that alcohol hurts people and society, and their solutions to the King Alcohol.[22] They were not alone. Mormon colonists in Utah came out strongly in favor of temperance as an outgrowth of their faith and added to the chorus against alcohol production, distribution, or consumption.[23]

Within this context, brewers found a difficult route to trod. They needed to clarify that their drinks were safe to consume in an era and region where water and spirits both posed threats. They needed to position beer as not alcoholic enough to be part of prohibition and, at the same time, produce high-enough alcohol content beer to make it enjoyable for those looking for a drink. How they performed the acrobatics is an integral part of the construction of beer terroir in terms of their political, social, as well as nutritional environment.

Healthy Water, Healthy Beer

Beer's nutritional healthfulness or harmfulness came from the minerals within water. Brewers were keen to point to the plethora of benefits bestowed upon their brews from pure, fresh, and clean Rocky Mountain water. Keeping water fresh, especially if one needs to haul and store it, was not the easiest task. For example, in 1884 Helena, Montana, a journalist discussed the difficulty in planning for water supplies, as people were too preoccupied with transporting water. Tired from filling, loading, and hauling kegs of water, it was a disappointment to all that the water quickly lost character. Instead, they noted that beer kept its flavor while in a keg, and therefore people forewent water for beer.[24] Similarly, in Neihart, Montana, an 1897 plan included adding a beer line to supplement the water supply, running beer through the same faucet. One direction for water, the other for beer.[25]

These comments may seem tongue-in-cheek and meant halfway flippantly, but they tap into a larger connection between water supply and beer consumption. Cities that navigated the issues of water supply outlined above still struggled to provide clean water. Beer, then, presented a more satisfying, palatable, and enjoyable option.

In some cases, however, it was a matter of health. Water carried many diseases, particularly typhoid during this period. A mix of public safety and advertising encouraged people to replace water with beer.[26] Added benefits included the caloric and nutritional contents of beer, which refreshed as well as nourished drinkers. The healthfulness of beer went beyond simply being better than water; in the time of growing temperance, brewers needed to show it was healthy and safe by its own measures. In this regard, Rocky Mountain brewers engaged directly with temperance forces who sought to stamp out intoxicating beverages.

Temperance snowballed to become a tremendous force in the late nineteenth and early twentieth century. Despite its lower alcohol content, beer still attracted the ire of teetotalers. Disputing a sliding scale of harmful effects from alcohol, instead they pushed for complete prohibition.[27] That beer was weaker than wine, which in turn was often weaker than spirituous liquors, did not dissuade some drys. To many in favor of temperance, all alcohol was corrupted, and thus any beverage, no matter how low the alcohol content, ought to be prohibited. One of the favored drinks of temperance was, unsurprisingly, water. However, as discussed above, water actually was not always the safest drink in terms of public health. It may not lead to drunkenness, violence, or other unruly actions that liquor did, but unclean drinking water did spread diarrhea, typhoid, and other diseases.[28]

Nevertheless, debates over the healthfulness of water and beer ranged across the Rocky Mountains. One Elko, Nevada, water proponent detailed the benefits of water over all other beverages—namely, flushing out our bodies.[29] In their opinion, shared nationally, the purer the water, the greater its cleansing potency. Not only was water beneficial on its own merits; beer was in absolute terms harmful. According to this view, it begets "rheumatism, local heart trouble, indigestion, painful swelling, and liver and kidney disorders."[30] The polemic here is illustrative of many temperance arguments.

Scientific data and arguments came to the aid of both sides of the water-versus-beer debate. The WCTU, a predominant temperance organization, organized informational sessions and pamphlets. A chapter in North Park, Colorado, posted a synopsis in 1910 about the supposed nutritional aspects of beer. As one might guess, they argued that beer in fact eviscerates the nutritional value of water and degrades the body's con-stitution.[31] The information came from a scientist, Alexander MacNicholl, in the American Society for the Study of Alcohol and Narcotics.

MacNicholl stated that water was a beneficial and pure drink by itself, but its intrinsic value becomes poisonous and tainted once mixed with alcohol or used in the creation of alcoholic drinks.[32] Further, the nutritional aspects of a pint of beer, MacNicholl argued, would not fill a teaspoon, and in fact, not only are the beneficial elements so small, but the manner of their mixture in beer actually makes them negative instead.[33] High-grade beers (those made with fresh ingredients and not with adjuncts or adul-terants) were not as harmful as low-quality beers, but clearly, the mes-sage was that beer was a harmful beverage to be avoided. The scientific and rhetorical work—however dubious, hairsplitting, or scientifically grounded—undertaken by MacNicholl reverberated across the US during the lead-up to Prohibition. Certainly, brewers and beer advocates did not agree with some of these arguments.

There are many ways water can absorb undesirable characteristics and, in turn, negatively impact the flavor of beer. There were two primary con-cerns brewers had with their water: mineral content and contamination. These issues, closely intertwined, stemmed primarily from runoff and pol-lution from mining and agricultural endeavors. Although they had limited power over these issues, they were able to take problems of water purity and turn them into a boon. This transformation from pollution to pure illuminates the way brewers across the region created, piece by piece, beer terroir of the Rocky Mountain West.

Mineral content and acidity of water significantly impact the flavor pro-file of beer.[34] Even trace amounts of certain minerals can lead to a completely different beverage. Thus, the quality and content of water matter a great deal. In large part, the minerals within beer derive from the water source; even today, different countries' water systems contribute significantly differ-ent levels and types of minerals to their beers.[35] Calcium, phosphate, mag-nesium, zinc, and other trace minerals within water systems are a handful of the many particulates and microbial matter pulled into bodies of water.

The minerals in beer are not a clear negative; they simply impart cer-tain flavors. Historically, beer styles gained much of their distinctive flavor from the local water sources' mineral consistency.[36] Famously, Eng-land's Burton-upon-Trent calcium sulfate content in their water is largely

attributed to the crispness and dryness of pale ales.[37] Today, brewers are
able to manipulate their water to a much greater degree than in past eras
if they desire to achieve a certain flavor profile.[38] Water's mineral content
makes modern re-creations of older styles of beer challenging, as records
for replication often exclude water minerality.[39]

In the Rockies, water from snowmelt carried with it into streams or groundwater the minerals found throughout the mountains. These same minerals
that drew miners and prospectors to the mountains infiltrated their water
systems and impacted the flavor of their beer. Much of the discussion thus
far has been about the natural impact of water—its scarcity and natural
contents—on beer. When Euro-American settlers came to the Rockies,
they dramatically manipulated the land-use patterns, which in turn fed rivers and streams a new diet of runoff.

Human-manipulated runoff posed the second obstacle. Mining, ranching, farming, timbering, and other activities that brought white settlers
to the Rockies severely damaged water systems. The effect of their poi-
soning was profound, and we still deal with the damage today, over a century later.[40] Brewers' water supply to make beer was polluted by the same
customers who came in for drinks after work. At once their raison d'être
and their largest obstacle, brewers faced significant challenges in promoting the quality of their beer to people who were largely responsible for its
contaminants.

Ranching, farming, and timbering all damage water cycles by changing
topography and landscapes. Reduction of native grasses and trees, additions of animal waste, and water requirements of animals and crops redi-
rect water streams and drain water basins. Groundcover slows water as it
rushes down mountainsides, and channeling this water in other directions
or reducing groundcover that can slow and absorb water results in floods,
erosion, and other negative environmental consequences. These influences
certainly impacted the quality and availability of water to brewers.

More detrimental than agricultural activities, mining most visibly
scarred the earth, drained water tables, and jettisoned by-products into
local streams and rivers. There has been much written about the effects of
mining on water systems, which can help provide useful commentary on

the beer terroir in the Rockies.[41] Mining, perhaps more than other activities, highlights the human impact on water quality. Brewers worked hard to position their beers as healthy, wholesome, and pure due to the purity of their water and cover up many of these negative environmental impacts.

Mining in the nineteenth century was a water-intensive activity. Sluices, hydraulic pumps, washes, and other steps involved water to help miners extract, clean, and refine ore. Wastewater from these activities, often now containing chemicals and heavy metals, was discharged into streams and washed away from the mining sites. The downstream impacts of these actions are well documented, as heavy metals and chemicals—especially mercury in silver mining—poisoned water systems.[42] As prospectors and miners eagerly tore into the earth to find their payloads, the waste tossed behind them was often out of sight and out of mind. However, these discharges did not disappear but trickled into groundwater that cities farther away pumped to drink and brew.

In the end, the truth was oftentimes the water was not fit to drink. Although a number of homemade strategies to test the quality and purity of water abounded, these were of dubious reliability. Scientifically driven filtration systems, such as those in the early twentieth century, were for beer factories and required substantial space, capital, and labor. One such filtration system advertised in a 1906 edition of *The Western Brewer* was twelve feet by thirteen to sixteen feet and constructed of bricks or concrete on a waterproof foundation. Beyond this, tremendous amounts of piping, coke, sand, gravel, and other requirements would put it firmly beyond the reach of most brewers.[43] With a grain of salt, beer proponents and antitemperance advocates did have a good point that untreated water was dangerous. One not unwisely stated, "Truth is that pure appearance, good taste and low temperature are no proof that water is fit to drink," and further, "A single drink of such water may bring on a severe case of typhoid fever."[44]

From Trouble to Terroir

So far, we have examined a number of obstacles and hindrances in brewing quality beer in the Rocky Mountains. Yet how did brewers overcome this, and in particular, how did they promote their beers as healthful drinks against narratives of temperance and realities of water pollution? In short, they claimed special considerations for their beer due to the pure Rocky

Mountain water from which it was brewed. Despite all the diseases, contaminants, and other impurities within their water supply, brewers clung to the concept of the purity of their water and its import for their beer terroir. Santa Fe brewer Charlie Neustadt argued in 1892, "The best beer in the country can be manufactured here largely because of the superiority of the water."[45] The ferocity in this defense was all the louder to overcome voices and evidence to the contrary.

The idea of the West being more healthful in general was not restricted to water quality. Americans flocked west in the late nineteenth century and into the modern day for its presumed healthful, rejuvenating qualities.[46] People ventured West initially for the dry, open air to cure tuberculosis and to return to nature and free themselves of urban enclaves.[47] Water's purity in the perceived idyllic West, unpolluted by urban factories and industries, did seem a much purer and cleaner variety. Brewers were keen to capitalize on this idea and built a beer terroir concept around this widespread belief.

In some cases, the purity and impurity of water for drinking and for brewing took quite a bit of doublethink and rhetorical work. The purity of water was crucial to brewers' success. They had to walk a close line between showing the advantage of beer over water and also portraying the water they used to brew as somehow superior and distinct. Water was a significant theater in the war to attract drinkers.

In 1910 Salt Lake City, a major brewery—Wagener's Brewery Co.—ran an advertisement column promoting their beer. In one breath they claimed Salt Lake City water was questionable and desirous. The water was typically of high quality and purity, but it was not immune to waterborne diseases. At the same time, they claimed their beer contained "distilled mountain water from our springs in Emigration canyon."[48] Ostensibly, boiling or distilling water for brewing or for drinking would rid it of many diseases, but in this brewer's opinion, it still mattered where the water came from. In direct defiance of temperance ideas, here beer enthusiasts posited that water was unhealthy until it became beer, not the other way around.

One Idahoan brewery ran a series of ads in the 1910s claiming, "It's not the water; It's the Beer."[49] This hints that their brewing process, again contrary to the above temperance position, actually cleans, amplifies, or otherwise enhances water's qualities when made into beer. Others claimed the purity of their water source and really got to the heart of beer terroir. Understandably, the pure mountain water was an oft-touted essential component of fine Rocky Mountain beers. George Mezger's Ely Brewery in 1888 brewed beer "from pure mountain water" and consequently had the best brew around.[50] His brewery was in Treasure City, Nevada, and another Nevada brewery, Carson Beer, in Butler claimed to brew only from "the purest water."[51]

There were a number of advertisements attesting to pure mountain water, especially in Nevada. Reasons for this stem from a variety of potential sources. Two in particular stand out. First, their water was particularly bad due to mining and other industries, and so the brewers overcorrected by portraying the purity and cleanliness of their water. Or secondly, in fact the water was particularly clean there, and to draw people to buy beer rather than drink water—an obstacle not met by brewers in towns with water in ill repute—they had to overpromote their beer's usage of the pure water. Likely it was a confluence of these reasons among others. As elsewhere, the brewers and marketers performed rhetorical labor to establish and promote their beer terroir.

One brewery waged a war against temperance and highlighted the narrow path brewers took to promote their beer against other beverages. They were not the only ones who attempted to navigate the changing beer landscape, but they provide an excellent example of the fluidity of beer and temperance. The Becker Brewing and Malting Company of Ogden, Utah, was on a mission to fight temperance. Becker's endeavor took on extra urgency in the 1910s as the nation, and especially the West, moved closer to Prohibition. One of the brewer's rhetorical two-front attacks targeted tea and coffee. These beverages were commonly promoted to replace alcohol. In contrast, Becker's beer pointed to the chemicals in tea and coffee, notably caffeine, that were similarly not ideal for the body. Beyond this, they promoted the healthfulness of their brews, stating it was "of pure mountain water" and as such "is of all beers the most nourishing."[52]

Again, here we see the value pure mountain water plays in promoting the healthfulness of beer over other drinks—in this case, tea and coffee. The irony here is that one action that makes beer healthier is replicated in tea and coffee: Boiling kills off many diseases in water. Nevertheless, it was the pure qualities of the water, coupled with the other nutritional aspects of beer, that made it a clear winner in Becker's mind. Interestingly, the next year, 1910, the Becker company seemed to join in with the temperance cause. This was a common move for brewers, who wanted to remain solvent and distance themselves from spirits and liquor.[53] Beer occupied a difficult middle ground between complete prohibition—backed by teetotalers—and drinking reduction, especially of high-alcohol drinks, which was more generally supported. They tried to align themselves as the temperate option to liquor and wines, which have higher alcohol content. True, the scorn of temperance largely lay with hard spirits, but the religious and ethnic nature of the movement also targeted wines (and thus Catholics) and beer (Germans and Eastern Europeans).

Becker placed themselves as the temperate choice: "The temperante [sic] man knows that Becker's Beer is the one and only beverage to be indulged in without injury to the system."[54] Even further, they were so bold to claim, "The placing of Becker's Beer on the market has done more for the cause of temperance [sic] than all the speech making and campaigning combined."[55] The reason for this switch may be due to an apt reading of the changing political tides. Whatever their reasoning, they continued to point out the temperate and nutritious nature of their beer due in part to "our pure mountain snow water."[56]

🍺

Water—access, availability, quality, and purity—is an immensely prized and crucial element in the Rocky Mountain West. Its role in many industries, such as mining and ranching, along with its political place between wets and drys meant that brewers had numerous obstacles to combat. They clung to the purity and nutritional value of the beer derived from their water sources at the same time they decried and castigated local water quality. This contradictory standpoint may seem merely an outgrowth of a shaky business enterprise trying to save itself—not a wholly unjustified stance. However, if we think beyond the mere immediacy of the businesses trying to hawk their products, we can see the emergence of a pan-alpine value of the water in their beer beyond a selling point. On the edges of these advertisements and debates emerges a sense of terroir in the water, that the mountain snowmelt that filled their vats and boiled their beer did have special properties. These properties—purity, cleanliness, and something external to eastern urbanized cities—were something unlike other places in the US.

4

History & Culture

Brewing and fermentable beverages are ubiquitous features of human societies. Brewing requires excess food, technology, high amounts of human and nonhuman energy, and specialized skills and tools along with stable communities in terms of population and migration. Communities do not need to be perfectly sedentary in order to produce fermented beverages, and seminomadic people have long produced fermented drinks.[1] This is especially true for much of North America, in the region that, through colonization and settlement, would become the United States.

Native American groups in the Rocky Mountain region—namely, the Utes, Arapahoes, and Cheyennes, among others—had fermented beverages as part of their foodways. There is a great deal of research, and continued practice, of brewing among Indigenous groups in the Americas. Chicha, a fermented beverage made from corn, is perhaps the most well known, but there are substantially more varieties. Just as beer is an umbrella term for a massive array of fermentation styles, Native American groups equally had a huge repertoire of fermented drinks.[2]

What will be discussed here are not these Native brewing practices but rather those imported along with white settler-colonists during the various Gold Rushes and land grabs in the late nineteenth century. These new practices, brought by Euro-Americans, obliterated much of the region's Indigenous landscapes and foodways and implanted their own. Over the decades, white settlers violently pushed out Native groups and peoples, forcing them onto marginal lands far from productive mineral or agricultural sites. This process was abhorrent and tremendously violent and laid the foundation of the story expressed here. White settlers from the United States and Europe imported their foodways into this region and re-created familiar spaces. Some of these spaces excluded nonwhites, and in many cases, it was illegal to share, provide, or sell beer to Native Americans.[3]

In their new home, these foodways and brewing traditions were not old and well-established. While cities and villages in Europe had deep-seated brewing traditions and calendric beer festivals, new townships in the Rocky Mountain West had none of these time-honored events or celebrations. As such, white brewers had to emphasize their local terroir in order to gain legitimacy and competitiveness in the face of older, more established, and likely more traditional brewers. They needed to validate their operations as well as conquer the foodways landscape. As there was no history or culture of brewing in the European fashion in the Rockies, brewers and pundits were especially eager to craft—and perhaps fabricate—the quality of the region for brewing ex nihilo.

Efforts to promote local beer as better than imported beer took center stage. In Idaho, Sile City Brewery produced "a finer and better beer than any shipped in here."[4] Further south in Carson City, Nevada, at Fisher's & Decker's Brewery, a tasting of beer from Connecticut resulted in all present agreeing that they could produce better beer than those brewed back east.[5] In Montana, one promoter, in an effort to encourage local purchases, said, "No better beer is manufactured in the east than Centennial, a home product of unquestioned quality and as pleasing to the taste as any made elsewhere."[6] Brewers' new Rocky Mountain homes provided them, in their estimation, higher quality resources and a finer end result.

Why the effort? Why did brewers seek to exert so much energy to sell products that ostensibly were better made in other cities? After all, if the region was indeed significantly better, then they would not need to broadcast it so vehemently over the decades. Conversely, if it was lower quality, then comparisons might invite opportunities to reveal inadequacies. Part of the answer lies in the same mentality as prospectors who staked out claims and trawled the mountains. By laying claim to beer terroir and making it part of their brewing atmosphere, brewers announced their ties to their new home and roots in tradition. In building a terroir, they plugged themselves into the stream of historical Euro-American brewing.

Beer is a deeply ingrained aspect of Western European culture. It is essential to many cultural groups across the world, but the specific brewing traditions of Western Europeans—namely, Germanic and English—became the mainstays in the US due to immigration patterns. This chapter will lay out the historical foodways that migrated from ancient Europe into the nineteenth-century American Rockies. Ultimately, this chapter explores how nineteenth-century Rocky Mountain West brewing was an outgrowth and continuation of demographic, brewing, and migration patterns in the United States. Lager as the predominant style connects many strains of American history.

American Brewing Origins

Europeans have been fermenting beverages for millennia. They are not unique in this aspect, but their brewing traditions do follow a particular trajectory. Wine today is often posited as a quintessential European beverage, but this was true only in certain times and places. Certainly, ancient and medieval Europeans had their own opinions on the qualities of wine versus beer, but the two fermentable beverages coexisted. Notably, historian Max Nelson discusses the attitudes toward beer of ancient and medieval Europeans in an aptly titled work *Barbarian's Beverage*.[7]

Brewing emerged from the ancient Fertile Crescent ages ago. Beer, unlike wine, required a multistep process to produce, and thus scholars believe it is unlikely to have emerged serendipitously.[8] Instead, it probably emerged in conjunction with bread making and city building between 8,000 and 6,000 BCE, although some push its onset back considerably to almost 12,000 BCE.[9] Growing alongside city centers, brewing as a craft and industry spread out across the world as people migrated in every direction.[10]

From Mesopotamia, brewing spread into Europe, first north and west into ancient Greece and, later, on to Rome. By the early medieval period, there was a long-established tradition of brewing in Europe. Most relevant here are deeply rooted brewing habits and foodways of Germanic peoples.[11] They brewed with their local agricultural and vegetable products and slowly over time codified formalized brewing styles. Brewing intersected, interconnected, and intertwined with local foodways, and people manipulated beer's ingredients and profile to suit tastes and needs.

Germanic brewers used various ingredients to produce their beers. One common early beer used *gruit*, an amalgamation of plant matter that scholars think primarily incorporated bog myrtle.[12] *Gruit* was part of a complex tax, legal, and territorial brewing industry in medieval Europe. Its addition was required as political leaders held rights to sell *gruit* and collect the tax. In many ways, this set the stage for hops' role in beer. The tax and taste structure in place made it easy to transfer from *gruit* to hops.

Although not immediately a central ingredient, over the centuries, hops became an essential part of German beer and supplanted *gruit*. In the eighth and ninth centuries, monastic brewing widened and accelerated this process. Monastic brewers began employing hops with greater consistency in their brews.[13] Hops became a fixed part of German brewing over time, and trading and interactions spread the gospel of hops throughout northern Europe. Its preservative and flavoring effects won over English consumers by the later Middle Ages.[14]

Primarily associated with cities, brewing grew as European tribes established larger and more numerous municipal communities. Settled and stable populations were ideal beer consumers in an era before refrigeration. Historian Judith Bennett has explored the development of English brewing from a small-scale cottage industry primarily practiced by women prior to 1300 CE to a male-dominated world that incorporated specialized labor and capital by 1600 CE.[15] Along with the growth of brewing came auxiliary occupations, guilds, tax reform, and even designated sections of cities.

Regulating beer served two purposes: tax and safety. Beer garnered a good deal of profits and thus was an important source of tax income. Due to the nature of their work, breweries catching fire was a common occurrence. Municipal leaders designated sections of the cities for breweries and bakeries to mitigate the risk of fire spreading to residential sections. This period also saw the most popular codification of brewing rules in fifteenth-century Bavaria, enshrined in the famous Reinheitsgebot.

By the 1800s, European brewing had evolved beyond a primary household chore that provided nutrition, hydration, or extra income. While this still occurred, major commercial brewhouses emerged designated solely for brewing.[16] They themselves may not have been particularly large, but they specialized in brewing instead of doing it alongside a set of other activities. It was this system that traveled across the Atlantic and colonized North America.

Brewing in the colonial and early republican United States produced English-style beers. The majority of colonists were from the British Isles, and they re-created familiar beers in their new home and imported ingredients and kegs from the mother country. English colonists in the American colonies imported British hops and malt to produce their beers;

nascent cottage malting and brewing developed over time, but commercial production—limited as it was—preferred British ingredients.

This changed rapidly during the American Revolution when English hop merchants cut off imports to the rebelling colonies. Over the eight-year conflict, American farmers sped up their production of local hops to take advantage of the sudden paucity of brewers' gold. However, after hostilities ended, trade resumed quickly and American growers once again faced imports that, though not as fresh and more expensive, were the preferred product for brewers. It would be another several decades before American-grown hops supplanted the majority of European-imported hops.

In the nineteenth century, American brewing experienced another burst of activity and growth. There were two major catalysts that morphed brewing into a new entity. First, industrialization with commensurate advances in technology, equipment, and productivity allowed some breweries to evolve from local brewpubs into beer factories. Second, in the US, a massive influx of Irish and German people in the middle of the century brought producers, consumers, and proponents of beer. Immigration changed the composition and nature of cities across the US. For example, in Chicago, Germans, Irish, and beer structurally changed the city's political, economic, and social climate.[17] Germans also brought a new style—lager—that would spread rapidly across the country, as it had in Europe. Today, lager is by far the most produced beer. In 2019, production amounted to almost 2 billion hectoliters, or 422 billion pints, quaffed in pubs across the world.[18]

Industrialization transformed European and American brewing. Louis Pasteur's 1864 discovery of pasteurization coupled with advances in refrigeration enabled brewers to prolong the shelf life of their products and market them to people much further afield.[19] Steam powered railroads that brought ingredients in larger numbers faster, and breweries harnessed that energy in their growing beer factories. Transporting people and beer, railroads utterly transformed the global beer industry. As railroads connected disparate fields and cities, space and time between them shrunk. Agricultural products, laborers, and consumers arrived in greater numbers at cities' beer factories, not necessarily in their homeland. While people moved from rural to urban enclaves, they also moved across regional, national, and continental boundaries.

Fleeing violence at home, Germans flocked to the US in massive numbers during the mid-1800s. Unlike many Irish who also traversed the

Atlantic during this time, German immigrants were more often wealthier, skilled workers with resources and abilities to establish breweries and re-create familiar scenes in their new home.[20] Germans built beer gardens to foster community and to establish social enclaves centered around beer consumption—typically lagers. This style experienced an inauspicious start in the US, but committed and industrious brewers—such as Coors, Anheuser-Busch, and others in the Midwest—built legacies producing lager beer.[21]

Lager is a particular style of beer that differs in a few ways from previously popular styles like porters or stouts. First, it employed a bottom-fermenting yeast rather than top, as in ale.[22] Second, it must be stored for two weeks after production. Lager derives from German and means "to store" or "to cellar."[23] This timeline favored capital-rich brewers who could afford to set the product aside for a period of time before selling it. The storage capacity required was immense, with some brewers cutting ice in the winter to store lager beer over the summer and contracting immense icehouses to keep the beer at ideal fermenting temperatures. This style stayed fresh for longer periods of time. Therefore, the confluence of industrialization and German immigration built the beer landscape in nineteenth-century America. These two forces went hand in hand across the US and spread out over the continent as white settlers moved and established towns and communities.

Rocky Mountain Lager

One of the places German settlers put down roots was in the American West. They built beer gardens and livelihoods under the shadows of the Rocky Mountains in Denver.[24] Like other white settlers, they spread out across the mountains in search of profit and prosperity. People had crisscrossed these peaks long before this time. Utes, Cheyennes, and Arapahoes traversed these ravines for millennia and continue to call this region home.[25] White settler-colonists also moved through the Rockies, following the 1849 California Gold Rush or in pursuit of landed stability through homesteading. People migrated throughout this region in pursuit of material prosperity and sustenance in culturally specific ways.

Whites did not begin settling the Rockies until the latter half of the 1800s. In 1858, a party of prospectors found gold in Cherry Creek—a stream that now runs through the heart of Denver.[26] Prospectors followed the stream into higher elevations, and masses of people came west looking for mineral wealth. The majority did not find their riches in extracting ore, but many found other lucrative avenues. Towns boomed as prospectors discovered rich veins of ore, and they busted just as quickly when initial surveys did not pan out. Over the next few decades, extractive industries marked the American West as people altered the landscape to capitalize on the natural bounty.

Communities often initially coalesced around mining claims. Around such a nucleus, stable communities eluded many formulations. If the town did not mature, usually when the mine dried up or proved unreliable, then the population moved on and abandoned their temporary homes. Unsustainable environmental and violent structures also disincentivized long-term settlement. Historian Elliot West tracked town development in the Rocky Mountains based on the permanency and status of local saloons. During a boomtown period, saloons were tents or wagons hastily constructed and quickly taken down to move on. If, however, the mine proved reliable and extra industries and businesses emerged, then saloons moved indoors into buildings as town builders constructed town squares and streets.

Many mines and towns in the Rockies did not move past this point. Even if mines were relatively reliable and people established other businesses in town, the constraints on many communities limited their growth. Supplies and food security in isolated towns supported only a limited number of people.[27] For many towns that remained as shells of their former selves, desires to reinvigorate the economy clashed with the economic capacity to do so. New breweries, due to their need for capital purchase, import of raw materials, and job-generating nature often were signs of better times ahead. Great Falls, Montana, fortunes dropped with the price of silver in 1893 but saw the establishment of a new brewery, alongside other industries, as placing their city "on the eve of a prodigious step forward."[28] Hope, then, came in a pint glass.

Occasionally, in prosperous towns, saloons became ornate, luxurious buildings that imported bars and mirrors and were frequented by elites.[29] This status often occurred in depot towns scattered throughout the mountains. Particularly, the Front Range housed many of these cities that gained their wealth sending raw materials—food, finished goods, mining supplies—into the mountains and receiving shipments of gold and other natural resources.

Figure 1: Here a few men gather to enjoy beer and other spirituous offerings from an established pre-Prohibition saloon. Note the large wooden bar and mirror, signs of a mature mining community. "Interior of Unidentified Saloon." Denver Public Libraries Special Collections. Call number Z-15707. https://digital.denverlibrary.org/nodes/view/ 1132144?keywords=%26ldquo%3BInterior+of+Unidentified+Saloon.%26rdquo%3B ++Z-15707.&type=all&highlights=WyJ6LTE1NzA3Iiwic2Fsb29uIiwiaW50ZXJpb3IiL CJ1bmlkZW50aWZpZWQiXQ%3D%3D.

Moving west in the nineteenth century drew Americans and recent immigrants in search of a new life. Gold Rushes pulled people from all corners of American society. Penniless laborers saw mining as an avenue to change their lots; business owners set up shop to supply workers with food, drink, and materials; the rich bought up parcels and claims and increased their wealth. In all, waves of thousands upon thousands of people crashed against the Front Range and spilled over the mountains. Many would recede in the coming years, but pools of people remained in the upper reaches.

Westward migration was tied directly to racism, American capitalism, and violence. Who could—or was allowed to—control western lands was a complex and racially informed competition.³⁰ The lure of a new life drew Americans into conflict with others seeking the same and the various Indigenous groups they sought to supplant. Wars between the United States and Plains Native Americans were violent, genocidal in nature, and occurred in serial waves that slowly—through various means of attrition—pushed many Native Americans away from prized lands.³¹ Continued immigration to the Rocky Mountain West led to commensurate conflict between the US military and Native Americans.

People came in droves. In the spring of 1859—the earliest massed caravans of people could arrange supplies, plan routes, and travel without major weather impediments—there were some one hundred thousand people trekking west across Kansas to the Rockies.³² This initial horde, along with subsequent masses, included a range of Euro-Americans. There was a significant international draw—particularly to Irish and Germans, who brought cultural values in their quest for material wealth. Over the ensuing decades of the nineteenth century, western states emerged and gained statehood as prospectors and others put down roots and built population centers.

To gain a sense of the population boom—and the beer-drinking demographics—examine table 1. This is the percentage of the modern state's population that was German. Note these are numbers of foreign-born people, not people who identified ethnically as Germans. For example, in 1890, Colorado, Montana, and Wyoming reported, respectively, 20 percent, 16 percent, and 15 percent native-born residents with one or both parents from Germany.³³ These stats hide the vast number of ethnic and cultural Germans who may have been born in the US but retained strong ties to their home country and moved to the Rockies. The actual number of people with strong German heritage, not to mention others from heavy beer-drinking cultures, added to the developing brewing industry's base. More so, it does not capture the wide backgrounds of other European immigrants who poured into the US from beer-drinking homelands.

Beyond all else, this table charts an explosive growth in western states and territories in overall population. These states experienced tremendous white population increases during this period. Though the German population as a percentage, with the expectation of 1890, was not an overwhelming number, every year thousands of ethnic Germans migrated into the Rocky Mountain region. Their population kept pace (and at times

Table 1: Rounded to nearest tenth percent

State	1860	1870	1880	1890
Colorado	1.6%	3.7%	17.6%	4.72%
Idaho	n/a	n/a	7.5%	3.2%
Montana	n/a	6.0%	6.3%	5.4%
New Mexico	0.6%	0.6%	14.8%	1.2%
Nevada	6.6%	5.1%	8.6%	4.7%
Wyoming	n/a	7.2%	14%	4.0%
Utah	0.4%	0.4%	2.0%	1.8%

Source: US Census Bureau, "1860 Census: Population of the United States" (1864), 565, 573, 578, 594; US Census Bureau, "Table VI: Special Nationalities 1870, by States and Territories," in *1870 Census*, vol. 1, *The Statistics of the Population of the United States* (1872), 336-39; US Census Bureau, "Remarks on the Statistics of the Nativities of the Population: 1880," in *1880 Census*, vol. 1, *Statistics of the Population of the United States*, 492-93; US Census Bureau, "State or Territory of Birth (Continued), Country of Birth, Foreign Parentage, Persons of School Age, Males of Militia and Voting Age, Conjugal Condition, Dwellings and Families, Indian Population, Alaska, Form of Schedule and Method of Tabulation," in *Eleventh Census*, vol. 1, pt. 1 and pt. 2, *Report on Population of the United States* (1895), cxlvi.

gained ground) with the onslaught of other immigrants from the US, Europe, Asia, and elsewhere. Once there, they set off in numerous pursuits to make money. One of which, enjoyed by many other recent immigrants, was brewing lager beer.

Temperance & Prohibition

Concurrent with colonization of the Rocky Mountains, temperance was a powerful force in the background of American society and politics. To be sure, Americans were drinking an immense amount of high-alcohol spirits and liquors. Americans drank incessantly; every occasion, no matter how small, throughout the day and late into the night warranted a drink.[34] Estimates range from 6.6 to 7 gallons of alcohol per capita between 1800 to 1830 to a more moderate (albeit still alarming) 2.5 gallons of spirits per capita in 1915, with additional gallons of wine and beer.[35] Valid concerns over domestic abuse, addiction, and social disarray abounded. In addition, there were less compassionate reasons for reducing drinking, such as industrialists worried about worker productivity, religious moralists worried about the soul of the nation, and racists who despised the drinking habits of recent immigrants.

In the West, these reasons for alcohol reduction echoed sentiments in eastern towns. Our vision of hardworking, dusty miners slugging nondescript but incredibly potent hard spirits was not necessarily true. Certainly there was a lot of heavy drinking and "treating" where the bartender paid for the first round to encourage drinking buddies to continue to buy each other subsequent rounds.[36] But this vision needs to be tempered with reports of mixed drinks, wines, beers, and other drinks that were more representative of drink choices in contemporary metropolitan settings.

National organizations such as the Women's Christian Temperance Movement and the Anti-Saloon League lobbied for prohibition laws on a state and national level. Victories in the early and mid-1800s were largely local ones; it was not until closer to 1900 that national prohibition began to gain serious traction.[37] Indeed, western states were the first to initiate statewide prohibition as a region, and the Rocky Mountain states had all enacted state-based prohibition prior to the passage of the Eighteenth Amendment.[38] The story explored here ends in 1921, with the (temporary) victory of Prohibition. The landscape of brewing after this moment changes dramatically, and the dry decade represents a palpable severance between brewing eras in American history.

In the waning decades of the nineteenth century, white immigrants from across the US and Europe moved en masse to the Rocky Mountain West, establishing familiar cultural spaces. Germans, who were more likely to have families, money, and specialized skills, imported their brewing traditions, values, and abilities. Beer brewing in the Rockies was not solely performed by German immigrants, but they were a large contingent of producers and consumers. The origins of beer terroir in the Rocky Mountain West began when gold-seeking white settlers built permanent residences and provided future brewers a consumer base to fuel their craft.

5

Barley & Corn

Atop Pike's Peak, Katherine Lee Bates faced east and surveyed the Front Range from the heights of the Rocky Mountains. On this perch, she drew inspiration for what would become a classic American song: "America the Beautiful."[1] She wrote this song in the 1890s, and it quickly became a patriotic anthem. One of the opening lines, "Amber waves of grain," was a testament to the agricultural hinterland that had developed in the late nineteenth-century American West. These fields of grain had not always been there but had been zealously plowed and planted to feed a burgeoning white settler population depleting native grasses and destroying Native American societies.

Grains, along with yeast, provide the fermentation side of brewing. Without grains and their sugars, fermentation would be impossible. Yet acquiring grains in the Rocky Mountain West was a difficult path. Population bloomed as mining rushes drew white people further into the canyons and crevices of the mountains. These rushes created quick camps, and those that sustained their momentum became towns and cities. No longer seeking to find quick riches, people cast about in order to secure food. Feeding these people with sufficient grain became a crucial problem to solve over and over. Early town boosters urged local farmers to plant grains widely and profusely. For their part, local farmers freed brewers from heavy reliance on imported grains.

Brewers in the American-European tradition did not want to use just any grain. Although most grains work for fermenting, lager brewers prefer barley over rye, wheat, and others. Preference for particular types of barley over other grains was cultural and practical. Culturally, barley connected white Rocky Mountain brewers to their English and Germanic brewing practices. Each type of grain has its own chemical makeup and requires different equipment to process, malt, and brew.[2]

Grain chosen and malted results in widely different character, flavor, and color. Two- and six-row barley (designated by the number of rows on the grains' head) were used in Europe to brew, and their properties were familiar to English and German immigrant brewers.[3] Cultural choices came at a cost. This preference curtailed the supply of grain for potential brews. In the mid-nineteenth century, brewers imported their grains and competed with others for this supply. Bakers, households, and others competed for grains coming into the West. Over the nineteenth and twentieth centuries, the issues brewers faced in their craft were the availability of barley and the constant temptation (and criticism) to use adulterants in brews.

Growing barley in the Rocky Mountains and building an adequate agricultural network to supply brewers were significant challenges. As more people immigrated and settled into the region in pursuit of mineral wealth or to profit off those who had it, feeding the populace grew as a concern. In Colorado, we can see this issue emerge as not only territory and later state boosters but also boosters for particular cities along the Front Range sought to attract settlers and thus increase their clout. A burgeoning town, with diverse industries, was critical to survive when the mines boomed and busted, and to encourage settlement, town boosters needed to promote agriculture.[4]

Growing grain required more than desire; farmers needed to adjust their rotations and techniques to adjust to the Rocky Mountains. This process took time and was not an overnight success. The mineral rushes took place largely during and after the 1860s, and local food production lagged behind the growth. Many preferred to strike it rich and go home than establish farms. This changed slowly, as camps became boom towns and evolved into cities and municipalities. The lag to grow barley to feed people as well as to brew caught up to the population over the later decades of the nineteenth century.

What this chart represents was a quick acceleration of barley production from 1860 to 1900. This rapid scaling up certainly benefited brewers, who over time developed a larger market from which to purchase barley. Availability of barley, then, was a challenge that lessened over time as various

Figure 2: A Coloradan barley field that would supply bakeries, breweries, and animals with grains and stubble. The height of the grain reaches to the people's chests. "Barley raised without irrigation, Skelton Ranch, Woodland Park, CO. Midland Ry (1900-1919)." Denver Public Library Special Collections. Call number MCC-1590. https://digital.denverlibrary.org/nodes/view/1019513. Used with permission from Denver Public Libraries.

foodstuffs became available. Table 3 shows the bushels of barley per capita. There was a peak during the 1880s and a second rise in 1900. This charts on to historical events, such as the Panic of 1893 when mines dried up and communities in the Rockies dissipated overnight.

Also portrayed in the chart is roughly how the states kept their barley production up with increasing population. What this meant for brewers was more grain as other agricultural products kept pace. As these states became more entrenched and established, the available barley and consumer base for brewers increased proportionally. Over time, farmers were more able to keep up with the rising population. As towns swelled over the latter decades of the nineteenth century, boosters' fears of consistent grain

Table 2: Production of barley in bushels (rounded to thousands)

	1850	1860	1870	1880	1890	1900
Colorado	n/a	n/a	35,000	107,000	832,000	531,000
Idaho	n/a	n/a	72,000	275,000	236,000	960,000
Montana	n/a	n/a	86,000	40,000	161,000	844,000
Nevada	n/a	2,000	295,000	513,000	237,000	224,000
New Mexico	5	6000	4,000	50,000	35,000	24,000
Utah	11,000	10,000	49,000	217,000	163,000	252,000
Wyoming	n/a	n/a	n/a	n/a	12,000	30,000
Rocky Mountain Total	11,005	18,000	541,000	1,202,000	1,676,000	2,865,000
National Totals	5,167,000	15,826,000	29,761,000	43,998,000	78,333,000	119,635,000

Source: "Agriculture in US Census Bureau 1870," https://www.census.gov/library/publications/1872/dec/1870e.html; "Statistics of Agriculture, 1900," US Census Bureau, p. 72, https://www.census.gov/prod/www/decennial.html.

Table 3: Bushels of barley per capita (rounded to thousands)

	1870	1880	1890	1900
Colorado	0.88	0.55	2.02	0.98
Idaho	4.80	8.33	2.81	5.93
Montana	4.10	1.03	1.22	3.47
Nevada	7.02	8.27	5.15	5.33
New Mexico	0.04	0.42	0.23	0.12
Utah	0.56	1.51	0.78	0.91
Wyoming	n/a	n/a	0.20	0.32

Source: "States and Territories," US Census Bureau, 1900.

shortages failed to materialize, though there certainly was not an overabundance. Today, some of the states that produce the most barley in the US are within the Rocky Mountain region. Idaho and Montana produce the vast majority in the region, followed by Colorado, Utah, and Wyoming.[5] Overwhelmingly, barley today is grown for alcohol production. Today, approximately 98 percent of barley is used for human direct consumption, with the remaining 2 percent along with spent grains from brewing and distilling for feeding livestock.[6]

Barley costs varied by state and time. Growing and harvesting grain, processing, and transporting all contributed to the cost to brewers. For example, in 1912, it cost one Idahoan farmer $2.75 per acre to harvest barley and another 11 cents to have it threshed.[7] Of this, they garnered seventy-five bushels per acre.[8] Part of the efficiency was due to new technologies and threshing machines that emerged in the early twentieth century.[9] Transporting them also impacted charges and increased prices for the brewers. In 1874 Nevada, transportation costs for barley were 2 cents per pound of barley.[10] At every step, the price from field to brewhouse increased. Yet this cost was a reduction from earlier in the century and resulted from expanding railroad lines and improved efficiency. Leland Stanford, a prominent western politician, recounted how in 1863 it cost $200 to $300 per ton to transport barley, or 10 to 15 cents per pound.[11] Brewers recouped some of their grain costs by reselling spent malt to farmers, who used the nutritious grains to feed their livestock.

Although homegrown barley was cheaper than products brought by rail or wagon from a distance, they often were not the same caliber as brewers had come to expect. The quality of grain was paramount in brewing. Once turned to malt, it provided the alcohol content, flavor, color, aroma, and other aspects of the final product. As such, brewers quarreled constantly over the quality of their beer by proclaiming the advantages the barley they chose imbued the malt they made (or purchased) and used.

Maltsters were tasked with turning barley into malt. This process was elaborate and skilled and posed a number of risks and challenges. Evidence for brewing in many cases was tied to malt production, and archaeologists have found evidence for malting in modern Germany as early as 500 BCE.[12] Some evidence suggests an even earlier date than this, and archaeologists continue to find evidence that pushes back the earliest known date for malting.[13]

Experienced maltsters were especially crucial in an era before heavy industrialization and standardization, when malting was done more by hand than machine. The period under study here was between these eras, where much was done by hand with the assistance of a growing number of industrial machinery.[14] Maltsters have an independent skill set from brewing but are an integral and highly skilled part of the brewing industry.[15]

In essence, malting is the process of preparing grain for fermentation. It is a multistep process, and its outcome determines a great deal of the color, aroma, alcohol content, and flavor of the beer. After harvest, the barley must be cleaned and stored in a dry, cool place. Grains selected for brewing are washed and then sent to a steep tank where the grain absorbs water. Ideally, barley selected has a low moisture level, around 13 percent, which during steeping increases to almost 50 percent. Once sufficiently waterlogged, the maltster drains the steeping tank and typically spreads barley out over a large area. The barley undergoes germination over the next several days.[16] Throughout, maltsters continually monitor temperatures, moisture, and oxygen to transform the grain into malt.

Once the barley begins to sprout, it undergoes a chemical process whereby it breaks down to provide energy to grow. This releases enzymes that in later stages of brewing facilitate fermentation. After turning the grains over for several days to ensure oxygen reaches the whole batch, the maltster truncates germination through kilning. Heating up the barley ends the germination process but preserves the released enzymes needed later in brewing.

The degree to which a maltster kilns their grain depends on the desired end product. By adjusting the heat and duration of kilning, maltsters can produce lighter malts for pale ales or darker malts for stouts and porters.[17]

In the Rocky Mountain region, there was a palpable transition from imported barley to locally produced. This changeover was significant, as the region became sustainable agriculturally but also brewers were able to develop their terroir and distinguish their region from others. Imported malt, either from other states or Europe, was held as the golden standard and continued to hold that position as Prohibition crushed the American malt industry.[18] Although much of this persisted, in the early years of the 1900s, brewers promoted their homegrown grain as the new high-water mark. They had become established in their industry and built a terroir around a crucial element of beer. Which malts were actually superior was entirely up to the perception of drinkers, but brewers definitely voiced their opinion.

Imported malt has consistently been held up as higher quality in the US, a situation that persists today. Brewers who wanted to produce true-to-form styles, especially German ones, often proudly reported their use of imported barley or malt to make their brews. Much of this bias toward imported malt was a terroir choice, one that reaffirms the superiority of certain ingredients, geographies, and practices.

Further, *imported* was a loaded term in terroir, particularly with beer. It can mean from a separate region, state, or country. While today, imports typically relate to international trade, in the nineteenth and early twentieth centuries, it had a more protective, narrow meaning. Any goods that came from beyond state or territory borders were imports, and some towns restricted it even further to mean anything from another city.

In early 1889, Valentin Blatz Brewing Company, a Milwaukee-based brewery, opened operations in Butte, Montana. They put imported California barley in their brews to create their Wiener beer.[19] The brewery in Milwaukee imported grain from California, and then the agent in Butte imported the beer from Milwaukee. Although we today may not see state-to-state transfers of ingredients or products as imports, reserving that world for international trade, the brewing industry at the time certainly would have.

From growers, brewers, maltsters, and consumers, import conveyed superiority—as in importing from a more wholesome or improved region elsewhere. After the passage of tariffs such as the McKinley Bill, one of the most significant issues for maltsters beyond price hikes was the higher quality of Canadian barley, as opposed to American, for malting.[20] In the Rocky Mountain region, import and terroir complicate the relationship between the land, agriculturalists, brewers, and consumers in interesting and complex ways.

The greatest boon to brewers came in the form of the McKinley Bill in 1890. It was an extremely restrictive international trade bill that instituted harsh tariffs for a massive array of imported goods, ostensibly to protect and nurture burgeoning but not quite stable American industries.[21] A protectionist act, it had far-ranging consequences throughout Europe, South America, and the Caribbean, which provided the US with raw and finished materials. This bill essentially cut off brewers from imported grains and forced them to rely on local sources of barley.

As was the case with many laws, this was a mixed blessing for brewers. Imported malt was held to be superior to local malts in many cases. The veracity of this may be questionable, but the importance here was the perception that it was indeed superior, that the terroir and maltsters from abroad were higher quality and better trained.[22] This status elevated the price of imported malt, adding to the built-in costs of transportation. With the imposition of higher tariffs, imported malt's cost diminished its viability to maltsters or brewers. In New York, for example, fears arose as maltsters who used to pay ten cents a bushel for grain from Canada now saw an increase to thirty cents.[23] Canada provided a great deal of cheap grain to maltsters who now found their supply cost prohibitive.

The impact of the McKinley Bill on malt in the Rocky Mountains cannot be overstated. Maltsters and brewers throughout the country sought out cheap, quality barley throughout the nation. Montana, for example, leaned into this and saw a boom in their barley sales to other regions in the nation. Proponents of the bill eagerly pointed to the superiority of the grain, while detractors suggested they were simply buying cheaper to meet demand.[24]

Reliant now on local grains, brewers had a new set of challenges. The bill made local grain cheaper by comparison, yet cheaper grain in the minds of drinkers could translate into a cheaper product. So much of terroir is perception and mentality, and if consumers fervently believed that German malt, for example, was superior to locally produced, it may diminish their perception of the taste of the beer. Recent studies have shown palatable differences between different types of malts, but the "imported" aspect that raised the taste profile is due more to mentalities than to malts.[25]

After the passage of this bill, there was a sea change in terroir. Now that imported malt from Europe was heavily restricted, growers and maltsters expanded their operations considerably. The desire for homegrown grain spurred a land-buying spree, which included land for barley. Chicago, a major brewing center in the US, notably needed to expand their hinterland to acquire the necessary barley for their operations.[26] In 1891, the year after the bill's passage, an amalgamation of Chicago companies bought 250,000 acres in North Dakota to supply their malting needs.[27] Although not directly in the Rocky Mountain region, it encouraged brewers to consider the quality of their grain and to develop an appreciation—real or portrayed—about the higher grade value of their grain.

Gallatin Valley, a region west of Bozeman, Montana, was one such region that latched on to the new dearth of grain and newfound terroir. Located along the northern Front Range, this region found itself in an ideal situation to provide not only grain but also a special Rocky Mountain version of barley. In 1893, the lagging barley sales across the nation could not satisfy brewers' and others' needs, and promoters thought Montana farmers should fill this gap.

One advertisement led off noting that maltsters were in desperate need of malt, though Bozeman's malting operations were in high working order. From here, they unabashedly announced the grain grown in the Gallatin Valley to be "declared by malsters [sic] to be the finest in the world" and further, humbly, that it was common knowledge the barley's quality was "absolutely unsurpassed."[28] Here, we see promoters for barley describing the higher quality of their grain based on their physical locations and advantages. The terroir specific to the Rocky Mountain region was responsible for this high-quality barley, which in turn resulted in finer malt and better beer. From this, a transition occurred where they exported barley terroir much in the same way they had previously imported Canadian or European terroir.

This turn toward appreciation of their homegrown barley required significant rhetorical work. Brewers, maltsters, and barley promoters had two tasks in front of them—one perception based, another rooted in realities on the ground. First, the brewing industry needed to promote and prove the quality of local barley and malt. Imported malt was firmly entrenched in consumers' minds as the key ingredient in high-quality, authentic, enjoyable beer. This was a mental hurdle to overcome, one that required advertising and a focus on the terroir that made local grain superior to imported malts.

Second, the issue of the quality of brewers' ingredients was a real, valid concern. Brewers could skip some of the malting process above and use unmalted barley. With an eye to cheapening costs, especially in locations with limited malting operations, this could maintain the integrity of the brew but cut out some of the skilled labor.[29] This often raised issues with maltsters, whose livelihoods were at stake.[30] This was especially true after the McKinley Bill, and fears of substandard grain producing substandard beer abounded.[31]

Beyond this, substandard brewing ingredients, called adulterants or adjuncts, weakened the beer, lessened the taste, and presented a beer not quite as advertised. In the worst case, adulterants, such as those found in liquor in nineteenth-century mining camps, could be used to boost the alcohol content of watered-down beverages. This was more typically a problem associated with spirits than beer, with descriptive names such as "Taos Lightning" to prepare potential imbibers for what they were about to consume.[32] Certainly, brewers took advantage of adulterants, and as temperance gathered strength toward the end of the 1800s into the early 1900s, people voiced concerns over harmful (or scary-sounding) additives. One writer in Nevada warned of substances such as glucose, boric acid, Indian hemp, bicarbonate of soda, and poisonous salicylic acid in beer.[33]

More often, substances were used as adjuncts that helped the beer ferment without using expensive barley.[34] One significant issue for brewers, and especially maltsters, was the quality of their barley and subsequent malt. In times of scarcity or leanness, brewers add any number of adjuncts or adulterants to their brews. Adding other grains extended their barley supply in order to produce more beer with fewer ingredients. While prudent and occasionally necessary, these actions brought on waves of discontent.

Primary adjuncts are rice, corn, or other cheaper grains or sugars that operate the same way barley does during fermentation but are cheaper to obtain. Although prohibited in a strict interpretation of Reinheitsgebot, not all brewers were fervent followers of this edict—at least not until barley became widely available. Also, not all brewers participated in German brewing conventions. Ethnic Germans constituted a large part of the white mountain-dwelling population, but large numbers of English, Irish, and other European cultural groups also inhabited this space.

Used extensively, adjuncts are not inherently a negative to beer. Adjuncts supplement, or supplant, barley in several ways. On a simple level, they contribute carbohydrates and sugars for the yeast to consume and produce alcohol.[35] They have always been a feature of breweries, and today their usage has exploded as major international brewing companies seek to continually beat the bottom line.

The disparagement directed toward adjuncts provided brewers an opportunity to promote the terroir of their beers and ingredients. The Fisher Brewery in Salt Lake City is one such example. A prominent brewery in the city, in the late 1880s, they posited terroir central to their high-quality beers. One advertisement claimed their "rapidly growing trade of [their] Lager Beer shows what can be done with Utah Barley, Imported Hops, and *No Substitutes* [emphasis in original]."[36] Excepting hops, discussed elsewhere, they tied their beer to both their high-quality local grain and the fact they eschewed any substitutes—that is, adjuncts.

Another region in Montana, the Gallatin Valley—mentioned earlier—particularly prided themselves on their grain. Above, we discussed a pundit's effort in 1893 to grow more barley in the valley, as it could not help but be top notch given all the natural advantages of this valley. Several years later, in 1897, farmers must have heeded the call and were proudly growing the best barley around.

The Centennial Brewing Company, a brewery local to the Gallatin Valley, attributed their high-quality beer to their local grain. Years after the passage of the McKinley Bill, maltsters were still dealing with a shortage of barley. Centennial Brewing used local grain and avoided the malignant effects of "corn and chemicals" used in eastern breweries.[37] Their beer was superior not only because of the absence of these adjuncts. Environs and farmers produced a natural and clearly superior barley. The rhetorical work here is not only that the terroir of their region positively impacts their beer but that they are also disparaging other regions' terroir. Eastern brewers, often meaning St. Louis, Chicago, or Milwaukee, were major competitive threats to Rocky Mountain brewers. In response, Rocky Mountain brewers attacked in multifaceted ways.

Brewers and maltsters keenly pointed to the benefits of local grain. Although production often lagged behind need, over time, the Rocky Mountain region developed an agricultural hinterland to feed fluctuating populations and to provide barley for brewers. Significantly, the 1890 McKinley Bill prompted intense investment in barley production and recentered terroir. Now that nearly all malt was locally produced, brewers pivoted from praising their imported malt to lauding the grain grown in their locale. As we have seen, Montana's Gallatin Valley certainly led the charge in this regard, but indubitably, brewers across the Rocky Mountain region posted the superiority of their beer due to their usage of the finest (i.e., local) ingredients.

6

Hops & Cones

Knowing, or fabricating, one's strengths necessarily needs to be counterbalanced by acknowledging real shortcomings. At some point, natural restrictions of climate, geography, and human effort forestall further advancements. Recognizing this point in its various manifestations is an important component of human geography. No matter how much one might attempt to cajole certain plants and agricultural products from the land, the requirements of the plant and the offerings of the land may not connect. Human history is rife with cataclysmic collisions between human aspirations or hubris and environmental constraints.

For Rocky Mountain brewers, this limitation comes at the intersection of hops and climate. Despite proclaiming the superiority of the various aspects of terroir—water, barley, and so on—hops perpetually flaunted Rocky Mountain growers' efforts at cultivation. Boosters encouraged growth, and indeed limited cultivation occurred. The hop cones produced, however, were laboriously grown. Simply, the growing conditions were ill-suited for the persnickety plant. Instead, brewers turned to other regions to supply their hop needs and loaned the terroir from them to add to their beers. In doing so, brewers understood their region as distinctly different—in this case, differentiated by climate constraints—from other regions in the US with easier and more vibrant hop production.

The US in general was not inept at growing hops, and in the period before Prohibition, the nation was a net exporter. For example, a report compiled by the United States Brewer's Association in 1909 showed that over a ten-year period (from 1899 to 1908), the US imported 47 million pounds of hops and exported 146 million.[1] Much of these hops came from

New England and the Pacific Northwest, regions with rich legacies of hop cultivation and production.[2]

Hops are a productive avenue into terroir. Recent examinations of beer terroir often center on hops. American hop growers in California and the Pacific Northwest along with others in Canada attest to the importance of terroir in hop development and flavoring.[3] Hops are a local product with tremendous variation and reflect even the slightest differences in growing conditions.

Hops maintain a significant hold on beer's branding and imagery, and it is fitting that much attention has been paid to the innumerable ways that hops differ field by field, variety by variety. This is one way this study distinguishes itself from earlier studies of beer terroir. This work seeks to incorporate many aspects of terroir and put them into conversation rather than hyperfocus on one element so as to understand the interplay and the multiplicity of beer flavoring sources. This chapter explores the history of hops in the US, efforts to cultivate them in the Rockies, and ultimately the strategies brewers employed to navigate their dearth of local hops. Terroir is not only a list of positives; it must also include determinants and obstacles that make regions unique and distinguishable.

When white settler-colonists first established communities in the US, brewing beer was an early concern. Procuring hops, however, proved a difficult endeavor. As colonies became more established, farmers planted familiar barley and hop varieties—there are native hop species in the Americas, but colonial brewers used hops that produced beers reminiscent of their homeland rather than their new environs. Occasionally, the strong tastes of American hops were a necessary evil to the costs of importing hops or when European hop gardens in the colonies failed to produce sufficient cones.[4]

Harvesting hops, like many harvest activities, was a flurry of activity for a short period of time. Extra hands to pick, sort, bale, and store hops were essential. The extra labor needed for these tasks, along with the specific skills and equipment needed, posed a barrier to entry for many farmers. Specialized farms and regions suited to hop growing in the US developed a profound infrastructure and system to ensure adequate hands, equipment, and buyers.[5] Family farms involved every member along with anyone from nearby towns and cities willing to work. Larger commercial farms in recent years require a balance of mechanical labor and technical agricultural skills.

How many hops are needed to brew beer? This depends entirely on the brewer and style. Today, there are various denominations for the amount of hops in a particular brew. IBUs—international bitterness units—roughly correspond to the amount of hops within a beer, as hops impart much of the bitterness in beers. In general, stouts and porters have the least hops, with pilsners and lagers containing a moderate amount. Famously, today, India pale ales (IPAs) contain higher amounts of hops, with adventurous brewers pushing beer's bitterness to the upper limits (and beyond) of what humans can actually taste.

Of course, this is a rough sketch, and late nineteenth- and early twentieth-century breweries offered a considerable array of beers. Predominantly, brewers created lagers, which contain a moderate amount of hops. Here, still, there was variation as brewers sought to stand out (or consider expenses) by adding more or less hops. Rocky Mountain brewers considered costs, freshness, and origin in the importing decisions.

The amount of hops brewers used, however, is fairly difficult to discern. Brewers protected their recipes, and smaller brewers likely did not write

Table 4: Hop production 1860-1920 (rounded to the nearest 100,000 in millions)

	1920	1910	1900	1890	1880	1870	1860
California	12.6	12	10.1	6.5	0.7	0.6	0
New York	0.7	8.6	17.3	20	21.6	17.5	9.7
Oregon	4.8	16.6	14.7	3.6	0.2	0.001	0
Washington	1.6	3.4	6.8	8.3	0.7	0.006	0
US Total	19.8	40.7	49.2	39.2	26.6	25.5	11

Source: "Fourteenth Census of the United States: 1920–Number 568 Bulletins Statistics for the States and Its Counties," US Census Department, 1919, https://www .census.gov/prod/www/decennial.html; "Twelfth Census of the United States–1900 Agriculture Part II–Crops and Irrigation," US Census Department, 1919, https://www .census.gov/prod/www/decennial.html; "Individual Crops," in *Fourteenth Census of the United States Taken in the Year 1920,* vol. 5, *Agriculture General Report and Analytical Tables* (US Census Department, 1922), 850.

them down to protect corporate and family secrets. However, we can esti-
mate based on an 1870 manufacturing report in Colorado. By no means
representative of all breweries across the region, it can nevertheless provide
guidelines for how many hops were needed. Averaging out the hop-to-
barley ratios, it shows an average of a 1.5:1 ratio.[6] A pound of barley, then,
warranted about a pound and a half of hops. This still leaves us unsure of
the total amount of hops used, but understandably, the amounts could be
quite high for large commercial brewers. More hop flavor and bitterness, of
course, requires higher ratios.

These hops did not come cheap. Although the Rocky Mountain region
developed a barley-producing agricultural system that provided its barley
needs, hops were not as amenable. These hops traversed significant distances, at commensurate costs. Beyond flavoring, hops were a key component of German and British brewing traditions, and the cost was a necessary
burden. Like malt, hops could be used in subsequent brews, albeit with less
flavoring, aromatic elements, and bitterness.

The Rockies were not suitable for large-scale hop production.[7]
Famously, Colorado's first batch of beer in 1859 was notably "innocent
of hops."[8] The dryness, soil consistency, humidity, sun, and rain patterns
were not amicable toward commercial hop cultivation. This, like several
other challenges facing Rocky Mountain brewers, many thought they
could overcome. One Nevada pundit posited that because rain affects
all plants alike and Nevada grows the best grain, they therefore should
grow the best hops.[9] Not considering the vast differences between hay
and hops, they further argued that "the few hops now grown in this State,
over bedroom windows and back porches are said to be far superior
to California hops."[10] A feeling perhaps shared by others, nevertheless,
hops grown here and there between clusters of homes would not provide
nearly the consistency, quality, or uniformity of hops desired and bought
dearly by brewers.

Farmers did pursue a dream of homegrown hops for a few decades in
the late nineteenth century. In the end, however, the capacity to grow hops
did not match the fervor. The 1880s and '90s were some of the highest-
producing years for the Rocky Mountain region's hops. Even then it never
reached the same level as either New England or the Pacific Northwest. In
1890, the Rocky Mountain states collectively wrested only twenty thousand pounds of hops from their fields.[11] Compare this to table 4 where, in

the same year, Oregon, the lowest producer of the states listed, harvested some 3.6 million pounds.

By 1920, none of the Rocky Mountain region states boasted hop production in the census records.[12] While there may have been some grown on garden patches and alongside houses or breweries, large-scale production was nonexistent. Two factors contributed to this: nature and politics. After decades of wrestling hop cones from weakened vines, farmers may have given up in favor of agricultural pursuits more amicable to their environment. Second, politically, by 1920 the beer and hop landscape had shifted considerably. In the 1910s, temperance gained significant strength, and national prohibition was enacted in 1920; many states enacted their own ahead of schedule and farmers saw the writing on the wall. Many western states initiated statewide prohibition several years ahead of the Eighteenth Amendment. Hops' main purpose is to brew beer, and if one's state was dry, farmers found little motivation to laboriously coax a few vines from the ground.

In essence, brewers borrowed terroir. Recognizing the advantage other regions had in growing hops, both in terms of economics and potency, they instead leaned into the imported aspects of their hops. As discussed elsewhere, the concept of import was broad and generously applied in this era. Certainly it could mean from another nation, and in the case of hops, it often did. The McKinley Bill of 1890 notwithstanding, imported hops continued to flow from Germany and England.

Rocky Mountain brewers' competition recognized the benefit of using imported hops in the beers. This was both a flavor brewing choice as much as an advertisement. Imported hops conveyed a premium, quality aspect to beer advertising. Particularly, this was true of German-heritage brewers. Anheuser-Busch, for example, offered a Bohemian Beer that was "King of all Bottled Beers."[13] This lineage owed its pedigree substantially to "the true flavor of choicest imported hops."[14] The American Agricultural Department pushed back against the connection between import and quality to limited success. In reality, American-grown hops often found their way to Europe. Yet perceptions of taste and terroir drew domestic brewers and drinkers to prefer the allure of imported hops.[15]

Hops from Europe faced a number of challenges before arriving on brewers' doorsteps. They mirrored domestic hopes, the difference being much greater distances. Anheuser-Busch's above advertisement was sure to point out that their imported hops were "preserved to a remarkable degree" so as to not lose flavor over their voyage.[16] The process of harvesting, cleaning, preparing, baling, and shipping hops was a labor- and technology-intensive effort. Keeping hops fresh over long periods of time proved another natural barrier.

European hops' cost prohibited widespread use in the Rocky Mountain West. Instead, hops from California featured widely in Rocky Mountain breweries. This may imply that they bought the majority from California, but there are other reasons why brewers paid so keen attention to California hop markets. First, though harvested in Oregon and Washington, hop bales directed toward the Rockies may have all come from San Francisco. A major trade port, it is feasible that hops from those states went south, then east. For example, in 1876, Oregon, Washington, and California collected nine thousand bales of hops in San Francisco to export out of the region.[17] A second reason was that California hop markets may have served as a type of barometer. Much as stock indices work in New York, California hop markets may have set the going rate for hops across the Pacific Northwest.

Newspapers zealously reported hop prices from both coasts. If we consider the geography, the Pacific Northwest would have been closer to brewers along the Colorado Basin, Western Slope, and perhaps the High Country. Conversely, New York hops may have been more obtainable by Front Range brewers and those on the eastern side of the High Country. Beyond this, rail access and sheer quantity connected the Rockies more closely to the East than West Coast. There were costs and advantages associated with both California- and New York–area hops.

Californian hop prices varied year to year and depended on local and global fluctuations in climate and economics. Bad years meant high prices. A glut in Europe would lower domestic prices, although these financial impacts were tempered by the costs of importing, especially after the McKinley Bill. Conversely, a weak harvest in Europe benefited domestic farmers, which in 1882 gave a huge boost to Oregonian hop producers.[18] Routinely, overproduction led farmers to bemoan their overladen hop vines. For example, farmers in the Seattle-Tacoma area grew a substantial surplus of hops in 1907 and drove prices down to eight to five cents per bale.[19] These prices were two cents per bale less than it cost to grow and package hop bales; the difference came from increased costs to grow along with an increase in local preference for imported hops from Bohemia.[20]

Figure 3: This image comes from a period after Prohibition but highlights the importance of good, consistent-quality hops. The man in the photo, Wheeler, is using a machine to test the hops for usage in Coors beer. "Adolf Coors-Wheeler Testing Hops (1948-1955)." Denver Public Library Special Collections. Call number Z-10223. https://digital.denverlibrary.org/nodes/view/1127564?keywords= number+Z-10223&type=all&highlights=WyJ6LTEwMjIzIiwibnVtYmVyIl0%3D&lsk= 655cda4f75e563c7a4dd008296c03e5d. Used with permission from Denver Public Libraries.

Although many brewers would have loved to buy bales of hops at eight cents per bale, that was not the case for Rocky Mountain brewers. A few advertised prices in the Rocky Mountain region reflect the price at the source and likely do not incorporate the cost of shipping the bales via rail cross country. Therefore, these prices should be interpreted as a baseline rather than the amount Rocky Mountain brewers paid on delivery. In 1888, English hop fields were damaged, and California hop prices in September of that year held steady at 12.5 cents per bale.[21] About ten years later in 1896, a surplus from California dropped prices sufficiently to encourage some growers to let their fields lie fallow to restore the price.[22]

In the 1910s, hop prices surged past these early benchmarks. As the hop market matured into the twentieth century, efforts were made to better organize and categorize hops. As such, in 1914, a hop growers association that included farmers in Oregon and California created a hierarchy of hops and attached minimum prices. Common hops, the lowest category, established a loose floor of twelve cents per pound, though it could fluctuate up or down a cent. The highest quality choice had a floor of fifteen cents and was not to exceed twenty-five cents.[23]

Obviously, with any attempt to fix prices, there are efforts by producers to undercut these established rates. This is exactly what happened the following year. A buying flurry emerged in the summer of 1915 as brewers sought to buy California hops. Although some vendors sold bales for either eight or nine cents, the prices were quickly rising to twelve cents purchased on-site.[24] Purchasing at location put Rocky Mountain brewers at a disadvantage.

Although California was a prominent location to buy hops from, it did not hold a monopoly. Brewers looked elsewhere for hops besides just California for better prices and flavors. One brewer in Pioche, Nevada, opened a brewery with a one-hundred-barrel-a-day capacity. Their flagship beer was a Boca beer that fermented for thirty days in one of their three 165,000-gallon storerooms. A substantial beer factory, they blended Californian and German hops, as they found the former too bitter when used alone.[25] Buying from multiple sources to achieve a specific hop profile highlights the ways they combined and amalgamated terroir of regions different from their own.

There was considerable reporting on hop prices and the market from New York, and so it was likely that California served as the ruler for hops on the other side of the continent.[26] The harvest of 1882 was a notable high-water mark where, per pound, hops rose to fifty cents. Due to mold in many fields, 1911 observers feared hops would again reach this height.[27]

In the nineteenth century, hop production in the Pacific Northwest—namely, sections of Northern California, Washington, and especially Oregon's Willamette Valley—took off exponentially. This pattern continues today, and the Pacific Northwest produces a third of the world's hops.[28] This region, however, was far indeed from Rocky Mountain brewers, particularly those in the High Country and Front Range. Despite the distance and the geographical barriers, these domestic hop fields were far closer than overseas farms.

7

Local Flair

An old joke begins as many old, good jokes do: A guy walks into a bar. After bellying up to the bar, he inquires who owns the joint. The bartender affirms he does. The guy then reveals he is an agent of a publishing house and asks if he could interest the bartender in purchasing some books, such as those by Shakespeare. The bartender replies that he has Wieland beer, Carson's beer, and other beers, but he has never heard of Jake's beer. The salesman tries to bridge this miscommunication and insists that no, he means a book. The bartender replies that in a few days, he might have some bock beer to offer as well. Exasperated, the salesman leaves in a huff, neither beer nor books exchanging hands.[1]

Besides a funny play on words, this joke highlights many of beer's naming conventions in the Rocky Mountain region. A handful of American breweries in the late nineteenth century were beginning to have named flagship beers. These named beers, however, were the exception rather than the rule, often only in cases of beer factories, such as in Anheuser-Busch's Budweiser.[2] Most commonly, particularly in the Rocky Mountain West, beers were named after the brewery—often the brewer's last name. This allowed people to connect beer to the brewer quite easily. This was the case with smaller brewers, such as the confusion around Jake's beer above, and with major producers, such as Coors.

Naming a beer after the brewer may seem strange, as today there is a surfeit of names for beers. However, even today, the major beer producers are named after their brewer/owner. For example, Coors, Schlitz, and Yuengling were all German brewers, named their breweries and flagship beers after themselves, began operations in the mid- to late nineteenth century, and today are major national and global beer producers. The name also helps convey information to drinkers. Those with British names built an expectation for offerings of ales, porters, or stouts. Likewise, Germanic-named breweries implied a selection of lager or bock.

Lager overwhelmingly was the beer of choice of brewers and drinkers in the US by this time. Americans had at times conversely enthusiastically and begrudgingly embraced lager imported with the millions of Germans from Europe over the mid- to late nineteenth century. Although ethnic tensions over alcohol were common, without a doubt, lager was the beer of choice across the nation.[3] This is not to deny the existence of other styles and varieties. One Colorado brewery in an 1864 advertisement listed champagne ale, cream ale, and "XXX" ale along with porter and stout from London.[4] The first three brews presumably were created in-house, likely in addition to lager.

This chapter will discuss the local variations and importance of local brewing—local in the sense of hyperlocal breweries. Although breweries across the Rocky Mountain West contributed to a collective terroir, they did so as individuals seeking to compete against breweries down the street, those a city over, larger ones in their state, and over time against beer factories in the Midwest. Local also in the sense of taking, adopting, or co-opting ingredients from the Rocky Mountain West. Brewers found themselves in a new environment but transplanted ideas and definitions of beer that were quite fluid.

Hyperlocal Beer

The Walters Co., a brewing company operating in 1890s Pueblo, Colorado, was aware of the challenges with distributing beer. Transporting beer was a risky business—not only because of crashes and lost beer but also since heat and environmental forces could spoil the whole batch. Shrewdly, they claimed their beer to be "the very best in the market."[5] Their claim's foundation lay in their distribution strategy, or lack thereof. Their local beer was superior to beers imported from Denver or further afield, since theirs was "not spoiled in the shipping as it isn't shipped at all."[6] Local production and consumption contributed to the superiority of beer terroir in an era when distribution across vast distances threatened to eradicate smaller local producers.

There are three distinct regions that were threats to breweries in the Rockies. First, geographically distant beer factories that, despite their distance, produced beers on a massive scale and drowned out competition through economies of scale. These are the eastern breweries that local boosters warned about. The next threat came from states within or neighboring the Rocky Mountain West. Hailing from the same beer terroir, breweries between Rocky Mountain states struggled to separate themselves from other state's breweries. Distance here still aided local production, but emerging metropolises over the nineteenth and twentieth centuries challenged smaller operations. The last threat came from breweries a town or so away. In this regard, brewers leaned hard on the ideas of local or hyperlocal consumption.

Local beer included not only the recipes but also resounding repetitious calls for locals to consume brewers' goods. This was not as straightforward as it may appear. Today, calls to drink locally are often craft brewers' response to beer factories. Because we have access to such a plethora of beer choices, the decision to drink something made in our vicinity is a conscious one rather than one made due to a paucity of options. Our view of the Rocky Mountain West drinking culture has been greatly informed by movies and media, which often do not reflect the historical reality. These towns, no matter their size, were not single-saloon, single-offering places. Rather, local producers were plentiful and there was a wide selection of alcohol from near and far.

To take one example, Leadville, Colorado, was a mining town at the top of the mountains with a local brewery—Gaw's Brewery. Some others came and went, but over time, into the late nineteenth century and early twentieth century, more and more beers were sold that were not made in Leadville. The expanding railroad system, along with improved transportation and refrigeration technology, meant that many of these mining towns that previously appeared isolated were in fact part of this much larger beer world. This connection was a boon to many, as it represented the continuation of their town's survival. For local brewers like Henry Gaw, this meant increased competition and beer wars that undercut and eventually ended his decades-long business. This was what Rocky Mountain brewers were against and why they fostered terroir in their advertisements to drink local beer.

Alongside advertisements for local beers came vociferous pleas to patronize local industry. Many brewers' claims to terroir were in response to other, larger corporations outside the Rocky Mountain West who also used ideas about the purity of their ingredients and importance of place in their advertisements. These statements became more threatening as beer factories in the Midwest expanded their operations and distribution. In the same way, calls to buy local rather than imported beer responded to encroaching beers from beer factories in the Midwest.

Over the late nineteenth century, brewers began to monopolize and conglomerate into behemoth beer factories. This often came at the expense of smaller breweries that were undercut in vicious beer wars and then assimilated into the major breweries. Though this often occurred in the Midwest near major beer towns such as Chicago, St. Louis, and Milwaukee, breweries in the Rocky Mountains were not immune.[7]

Figure 4: Housing was sporadic and often required people to cram into small spaces to conserve heat, fuel, and money. This group of men were likely laborers in extractive industries and patrons of local breweries, as two of the men in the front attest. "A Woman and Her Boarders (1890-1910)." History Colorado. Accession number 84.193.162; formerly F38,016. Handwritten on envelope "Mining-Miners-Homes." Call number CHS.X4876. Used with permission from History Colorado.

In many ways, calls to consume locally reflect the cleanest interpretation of beer terroir in the Rockies. By definition, they argued, their local beer was superior to imported beer, no matter how far away it traveled. This was a stout defense against consumers drinking imported beers from other Rocky Mountain cities or states farther away. It also, in a real, practical sense, reflected the challenges of transporting beer. Refrigerated railroad cars had been in service soon after the Civil War, though they were slow to move west. As such, high heat during the summer months posed no shortage of difficulties in keeping beer cold with ice and insulation as it traveled city to city.

More so, accidents happened in the Rocky Mountains. Driving through mountain roads at any time of year increased the chances of accidents and spills. For example, a Fourth of July celebration was dampened in 1904 Coal Creek when a carload of beer spoiled along the way. Another car was sent up in its place, likely at cost to the brewery to replace the spoiled batch. The entire spoiled contents were dumped; it was not fiscally viable to return the spoiled beer to the brewery for destruction. Not all was lost—local animals were able to recover some of the lost calories. Hogs roaming the nearby area smelled the beer and went to partake. They reportedly "drank freely of the spoiled beverage and became intoxicated."[8] One sow "indulged too freely" and got stuck in a pool of beer.[9] The July heat ruined the first carload, and though this had a humorous ending, not every beer transportation accident was as innocent.

Accidents could also occur in city transportation. In 1910, a beer wagon collided with a car in Ogden, Utah. Casualties included the wagon, car, and some three hundred bottles of beer.[10] Collisions such as this were not infrequent and resulted in destroyed vehicles, shattered glass, and a great deal of spilled liquid. Transportation was a financial risk breweries incurred, one that resulted frequently in lost product.

These calls responded to real threats to breweries' stability and solvency. The appeals to home loyalty and local superiority were natural consequences of previous statements supporting terroir. Breweries and boosters expended a great deal of energy constructing a bulwark defending the benefits of local ingredients for local beers. Brewers aimed to increase consumption of local beers, narrowly defined, rather than beers imported from other cities or other states.

Calls to patronize local breweries rang out as soon as breweries barreled their first batches of beer. Some leaned into this more than others, either through target advertising or other tactics. For example, one brewery in 1874 Wyoming was simply called Home Brewery and brewed, sold, and delivered bottled lager.[11] An emphatic and clear call for beer drinkers to consume local rather than imported beer comes from Anaconda, Montana, in the 1890s. A full column called "About the City" listed goings-on, which was common in newspapers during this period. What stands out are, in one day, the repeated calls to protect Anaconda's brewing industry.

As virulent and threatening as interstate beer production and distribution were, closer rivalries often were more dangerous. At the turn of the century, a new brewery in Reno was the object of much consternation in Carson City, a town separated by slightly over thirty miles. Carson City town defenders minimized the danger posed but also issued a call to action for "the people to take action in the matter for self-protection."[12] Although buttressed by the "superior quality of wet goods" and high wages, there were nevertheless calls to the community saying, "The home people should look to her own people first."[13] Clearly, a distance of some miles and ostensibly superior beer were not enough to insulate Carson breweries.

The column is essentially a list of blurbs, containing news and ads. They promote, "Call for Anaconda beer and take no other." Not only should residents "drink Anaconda beer and keep the money at home"; they ought to also "patronize home industry and drink Anaconda beer." These exhortations ended with, "The only home industry in the line of beer is the Anaconda Brewing company."[14] Within several inches, a reader scanning the column would have been hit by these messages over and over. The urgency, frequency, and consistency of these messages make clear the threat not drinking at home posed for local brewers as they sought to fend off incursions from brewers. For example, just a few years later, in 1899, Anheuser-Busch promoted their bock beer in Anaconda as the "best in the world."[15] Competing with major breweries, who could provide consistent beer at low prices, brewers appealed to locals' pride in their neighborhood institutions. In many places in the Rocky Mountain West, once major beer factories began advertising and listing selling agents in the local newspaper, the fight was on for local breweries' survival.

Brewers and pundits in other towns and states in the Rocky Mountain region, and indeed across the US, echoed these same sentiments from

Anaconda. Like other aspects of terroir, Rocky Mountain brewers recognized their competition and mirrored tactics other, external brewers
employed. Often, this was in response to major, national breweries such
as Anheuser-Busch, Schlitz, or Pabst, but towns resisted "imported" beers
from any competitor, especially within states as neighboring towns' brewers fought to remain solvent.

The home industry complicates the nature of imported beers. This was
not merely a domestic term for the entire US. Nor was it a term that even
covered a single region at times. What counted as "home industry" could
include a single state with every beer produced outside the state or territory lines excluded—even if only a single state over.[16] Brewers themselves
narrowed ideas of "local" to be inclusive of city limits, and they labeled any
beer from any municipality or community outside these borders as failing
the home-industry standard. In doing so, they crafted the terroir of their
locale in relation to the region. Most notably, Denver was home to major
beer factories, relative to other breweries in the Rockies. Zang's Brewery,
Rocky Mountain Brewery, and nearby Coors in Golden produced a great
deal of beer between 1890 and Prohibition. One of the larger breweries in
the Denver area in 1892, Zang's Brewery, contained a brew tank capable
of holding four hundred gallons at a time and produced 150,000 barrels
in 1892.[17]

Production levels of this enormity towered over breweries deeper
within the mountains. Breweries along the Front Range had better access to
resources—consumer base, fuel, equipment, ingredients—that advantaged
them over brewers in the Great Basin, Western Slope, or High Country.
For example, Dillion Brewery in Butte, Montana, produced two thousand
barrels in 1903, and in 1902, a local brewery in Albuquerque, New Mexico,
produced forty thousand barrels.[18] These ranges still differed greatly, and
one can see why smaller brewers felt threatened by neighbors all the while
Front Range breweries threatened to overwhelm and replace them all.

"Denver beers" encapsulated the number of beer factories housed
in Denver's vicinity. Interstate and intercity rivalries boomed in the beer
industry and were extenuations of other rivalries—particularly in the West,
where boom towns or cities competed to ensure they attracted the most
income to remain solvent. For example, along the Front Range, Denver
and Pueblo were significant rivals. They competed to be the dominant city
of the Front Range to receive raw materials from the High Country and

forward them on to eastern manufacturers. Depot cities' character enabled the development of large breweries, which in turn exported their products broadly beyond the city limits.

Pueblo breweries issued a series of advertisements decrying Denver beers. One suggested the upstanding Pueblo brewers used only the highest quality hops and malt, whereas Denver brewers opted for "glycerine and molasses."[19] This was not the only negative review of Denver beers, or even the most outlandish. At the same time, there were numerous promotions in favor of Denver beer. Indeed, the day before aspersions of glycerin and molasses were voiced, the *Colorado Daily Chieftain*—the Pueblo newspaper that issued these same denigrations—ran an ad extolling the virtues of Denver beer. The message: "Everybody takes it; everybody likes it."[20] Battles between cities and states abounded, with newspapers fanning the flames and playing both sides.

The methods with which cities and towns differentiated beer provide clues to the way people in the Rocky Mountains viewed not only their competition but also regions. Gresham and Patterson was an alcohol dealer in Cerrillos, New Mexico. They ran a series of ads promoting their offerings. Typically detailing generic wines and liquors, one ad specified their beer options. Promoting their "Fresh Rocky Mountain and Denver Beer" signifies beer terroir and clarifies different regions or terroirs within the great Rocky Mountain West.[21] Beer, a manufactured product, contained worlds of political, economic, social, and existential tension.

Spruce

Brewers latched on to the capacity for individual experimentation in the unregulated brewing industry in the nineteenth-century American West. Recipes for beer made from ginger, corn, molasses, and spruce, among others, abounded. Many of these, such as corn or molasses, were used as adjuncts in creating staple beers—meaning they were major, but not primary, ingredients in brewing. If proper grains were in short supply, increasing adjuncts produced equal quantities of beer, if not identical quality. San Juan Brewery in Del Norte, Colorado, claimed their beers were "entirely free from adulteration" and were "pure in every aspect."[22] In producing pure, adjunct-free beers, their "heavy sales [were] proof of its superiority."[23] In a pinch, these ingredients could facilitate the creation of beer. Brewers brought these strategies with them as they made their way in their new environment with uncertain supplies.

Brewers and drinkers wanted a consistent, palatable, comfortable drink that was not expensive to brew or purchase. Brewers, then, walked a fine line between using local ingredients and experimenting with nonbarley bases and driving consumers away. Their local beers showcased the ways in which brewers embraced their new location and used it to attract rather than divert potential customers.

Some of these recipes push and extend beyond the boundaries of the modern definition of beer. Beer today is a beverage that requires certain ingredients and follows certain technical steps. Yet today and in the past, it is a flexible, malleable, and widely encompassing term. It could mean more than a beverage with malt, hops, yeast, and so on; it could also mean alcoholic beverages that contained less alcohol than wine but were still intoxicating drinks. One recipe, indicative of many self-guided brewings, called for only molasses, ginger, and water: Mix a half-gallon of molasses and one spoon of ground ginger in five gallons of water, and after a few days, it results in "good beer."[24] Brewers and evaluating entities today may question the validity of that term. In the Rocky Mountain West, as elsewhere in the late nineteenth and early twentieth centuries, this was a completely legitimate moniker.

Corn in particular was a common adjunct. It continues to be used extensively and occasionally is vociferously attacked as lessening or diminishing beers. Notably, Coors uses corn and corn syrup in their beers, something that Budweiser continually hammers and broadcasts.[25] Although typically, corn was a secondary ingredient, there were a few instances where it was the titular ingredient in beers. Biases against corn beer as opposed to more familiar lager or ale worked against the beverage. Also, corn was needed for distilling, and the crop was directed toward distilling whiskey or bourbon—a more lucrative use of corn.

There are a few categories of recorded beer types that defy typical beer styles. Lager, although dominant in the region and the nation, was not challenged by these local brews. These are tremendous avenues to see the resource use and the ways in which brewers explore their terroir and craft a local taste. All the while, brewers upheld traditional brewing styles and patterns. Beer offerings arose from the confluence of experimentation, flexibility, and tradition.

Specifically, this section will focus on spruce beer, a beverage at once common and understated. This was a local drink that was common in the US, essentially brewed anywhere there were spruce-bearing trees. Here we will focus on this beverage, as its highly localized nature provides a great sense of terroir. In its hyperlocalized form, spruce beer also can easily be inscribed with meanings and values of the community, which makes it a flexible, malleable, and effective beer for messaging.

In many ways, spruce beer recipes resembled the molasses beer recipe above. It involved an extra ingredient but the process was similar. Little mashing, malting, or other standard brewing practices were followed. How, then, can we classify spruce beer as beer, if people made a fermentable concoction, let it stand, and imbibed it? What makes this a beer rather than low-quality alcohol?

Beer has always been a tremendously fluid beverage. People introduce any number of ingredients, steps, equipment, values, or meanings into the brewing process. While spruce beer and other local recipes do not contain malt or a mashing period, this should not disqualify them from categorization as beer. First, beer is a nebulous, amorphous beverage at best and to delineate its boundaries does more harm than good. Beer adapts and evolves in every generation, and to truncate it and place it firmly within historical confines reduces our ability to examine its past and explore the nuances contained within every pint, keg, and barrel.

Of course, this wide framing and broad understanding of beer may experience pushback, and strict beer styles exist for a multitude of reasons. It is one aim of this work to explore how beer exists as a discrete product and also as a nebulous messenger and communicator of environmental, ethnic, and cultural values. In my opinion, expanding our views of what constitutes beer allows us to widen our historical gaze and incorporate a wider range of fermentable traditions.

Second, this work is concerned with perception, which furthers the ambiguity and openness of beer. To take a provincial view of not only beer but also people's definitions of it would discard a great amount of people's lived experiences. Certainly, extending the definition of beer indefinitely is equally harmful. Here, the analysis will include only beverages that people at the time considered beer, designated by their own terminology. While their beers may fail many prerequisites we have for beers today, in another one hundred years, historians may look at current brews, such as milkshake

IPAs, that test or surpass even enlarged definitions of beer and consider these brews unbeer-like. This work considers the realities and perceptions at the time, and for many, spruce beer was beer.

Spruce beer, as its name implies, primarily contains spruce needles and stems. These needles replaced hops in their bittering, preservative, and flavoring profile.[26] Like most beers, this provided numerous health benefits to drinkers. Famously, Captain Cook had spruce beer brewed on board in order to prevent the onset of scurvy.[27] In Europe there was a slight presence, brewed from the Norway Spruce, but in the United States and Canada, spruce beer became a household staple.

It was a favorite beverage of frontier colonists or people on the move, such as explorers or soldiers. As colonial armies or prospectors trudged along the landscape, spruce was readily available to hydrate, nourish, and please these roving bands.[28] In the Rocky Mountains, spruce trees were especially abundant and therefore a ready resource for brewers seeking to expand their offerings, tap into the terroir of their newfound homes, or provide a cheap beer already ingrained in American beer-drinking culture.

On the surface, spruce seems a random ingredient for brewing. After all, there are a great variety of vegetable matters that are fermentable. Spruce, however, has long been an ingredient in the culinary history of settlers and colonizers of the United States, who themselves likely imported its usage from Northern European practices. Although not explicitly preferred over other styles, a familiar beer was better than no beer at all. Early white settlers, particularly in Canadian Nova Scotia and Newfoundland, drank copious amounts of spruce beer in part for its nutritious elements.[29] Occasionally mixed with rum to create a separate drink, from its early days, American spruce beer occupied a liminal space as a healthy, temperate drink as well as one used for unhealthy intoxication. The Canadian and American black spruce commonly supplied brewers with the twigs, stems, cones, and needles used to prepare the beer.[30]

We have a plurality of historical spruce beer recipes that were shared and preserved. How-to guides for home production abounded in the eighteenth and nineteenth centuries. They were not often commercially produced on the same scale as lager. As such, the recipes were not guarded in the same way, and recorded recipes in bartender manuals, brewers' handbooks, newspapers, and all other mediums provide a wealth of spruce beer varieties and styles. This beer was for everyone, and everyone had their own certain twists on their brew.

One 1871 Ohio recipe highlights the cottage-industry aspect of spruce beer as well as gender divisions in labor. It was penned for the housewife or "well brought up daughters [sic]" to know how to make a wholesome, temperate beverage.[31] In three gallons of water, mix in one-and-a-half pints of molasses, add a tablespoon of "essence of spruce," and finish off with another tablespoon of ginger. Add a gill of yeast, which was about a cup of yeast previously used in brewing beer, and let stand for a day.[32] After this, home brewers obtained a slightly alcoholic beverage brimming with nutrition and flavor.

This recipe—simple, quick, and cheap—reverberated across American cities. Word for word, this recipe was printed in newspapers in disparate states such as Wisconsin and Tennessee and the basic format copied in New Jersey and Michigan, among other states.[33] Other printed recipes contained a great variety of ingredients, some more reserved than others. A Michigan recipe recommended adding a lemon peel.[34] A more involved recipe from early twentieth-century Washington, DC, included oil of wintergreen, oil of sassafras, and sugar in addition to spruce and yeast.[35] In the Rocky Mountain region, a Utah recipe suggested hops, sassafras root, ginger, and allspice in addition to spruce and molasses.[36] These alternatives fail to capture the entirety of the range of spruce beers, and likely people adjusted the recipe in any number of ways. Adding different plant matter, spices, or other additives could tremendously change the constitution, taste profile, and congeniality of the brew.

Primarily produced, consumed, and appreciated on a local level, there were some named spruce beers. To call them commercially made is not quite accurate. There were spruce beers made, named, and sold in stores but not as a primary industry—meaning a brewer would predominantly rely on sales of ale or lager to support their business, with spruce beer supplementing their income. Few, if any, relied completely on spruce beer, as its easy replicability made it financially unstable. The nature, history, and commercial viability as a large-scale product disincentivized brewers from producing spruce beer in the same manner they did lager. Rather than designated breweries producing it as a commercial activity, spruce beer operated more in the sense of being sold as surplus or by local artisans. For example, Fox's General Store offered Gruman's spruce beer in 1877 Colorado.[37] In Idaho,

E. Lee in 1863 was the proprietor of the Spruce Beer Shop, undoubtedly a good place to find its titular beverage.[38]

It was likely that people who bought and sold spruce beer had favorite brewers or vendors, even if today it is difficult to discern where retailers bought their supply. Or, as in the case of many breweries, saloons and vendors may have had their own in-house operations.

Spruce & Temperance

Another benefit of spruce beer was its ambiguous stance for drys—that is, people in favor of prohibition. Initially, early supporters in the eighteenth century enjoyed it for its low alcohol content, particularly among soldiers and sailors. Similar to ginger ale or beer today, it was a brewed beverage that did not necessarily have a high or present alcohol content. Military and naval leaders provided large amounts of spruce beer to replace spirits, provide nutrients and calories, and combat scurvy. They could be confident that the brew would not lead to intoxication, violence, and disorder in the same way spirits did due to the lower alcohol content. This allowed brewers to skirt a number of temperance complaints, especially as states enacted local prohibitions.

Temperate beers, such as the Ohio recipe above, were another way we can use spruce beer to expand conceptions of beer. Spruce beer "free from alcoholic influence" allowed brewers to create this beverage while skirting around many temperance laws coming into effect.[39] The American West states enacted temperance far ahead of Prohibition, and those vending spruce beer or recipes needed to be careful to note the beer's temperate nature.

Spruce beer in the Rocky Mountain region certainly fits the political and alcoholic landscape. Though difficult to know exactly how low (or high) the alcohol by volume was in certain spruce beers, its temperate nature was touted extensively. This does not mean necessarily that it was alcohol-free, although that certainly was the case at times. Rather, the beer was a more temperate option than spirits, wine, and even other beers. Low enough alcohol to fly under the radar, it still had vestiges of alcohol and thus formed a gray area in the march to Prohibition.

Many spruce beer promoters advocated the beverage in much the same way lager beer drinkers did. They touted its healthfulness, wholesomeness, and flavor. This facilitated conversations about spruce beer as the temperate option, and it often was promoted as such. For example, in 1889, Eugene Katzenstein began selling his spruce beer in San Marcial, New Mexico.[40] Their base of operations was in Socorro, a town about thirty miles north. The commercial viability of selling spruce beer in other towns hints at the success of Eugene's recipe and business acumen. One aspect of this business plan was to sell "spruce beer and other temperance drinks"—a strong strategy.[41]

New Mexican beverage pundits favored low-alcoholic drinks. In some regards, spruce beer as a temperate option supplanted other regional fermented drinks. Pulque, also made from tree products, however, was not received as favorably as spruce beer. Its flavoring postfermentation was notably abrasive for new consumers unaccustomed to the beverage. One commentator noted after fermentation and the "alcoholic principle is fully developed in the liquid it is not only extremely intoxicating, but has a most abominably putrid smell."[42] The alcohol content did not endear this imbiber, but more so smell was enough to put them off. The author here perceived a much stronger negative taste from fermentation than their peers. Perception was not only about how terroir makes a beer better than others; it was just as much about how the process can diminish the quality of the drink. It was only after fermentation that the rancid taste appeared, prior to this, the flavoring was reminiscent of spruce. In this way, drys tried to steer others toward spruce beer instead of stronger drinks to sate their appetites.

Spruce beer was central in debates in New Mexico between drys and wets. One wet pundit jabbed at the drys in a short skit pitting a fictitious, aptly named Reverend Coldwater of Drytown against an impish boy. Coldwater—a stand-in for drys—encourages the boy to convince his parents to try spruce beer and to "stop drinking that vile stuff." To which the boy retorts, "It'll be a cold day" when boys rush around for spruce beer.[43] New Mexico was not the only spruce beer battleground and is emblematic of other arenas.

The beer landscape in the late nineteenth and early twentieth centuries was dominated by lager. Brewers in the Rocky Mountain region produced this beer en masse. Yet there remained on the peripheries an important market for local beers made with readily accessible ingredients. These beers, of which spruce was the most popular, offer a window into the cottage-industry aspect of brewing. Made primarily for home or proximate consumption, these beers contained a great deal of meaning and value for a host of reasons. Drys enjoyed their low alcohol, wets appreciated their legal-temperance ambiguity, and consumers appreciated their flavor and flexibility. Brewing drew brewers intimately into their hyperlocal economic and natural environs. In doing so, brewers and consumers built an appreciation for their place and manifested their views into regional conceptions of terroir.

8

Drinking Beer, Tasting Terroir

The return on investment from marketing campaigns is a crucial question posed by any business. How many dollars spent will result in an increase in revenue? What is the trade-off? At what point has a company saturated the advertising space and no longer needs to build brand recognition? This is an especially murky and unclear issue in regard to beer terroir in the pre-Prohibition Rocky Mountain West where brewers attached beer terroir to every possible element of brewing in an attempt to establish market share. Ultimately, did consumers buy the beer terroir, or did consumers just buy beer to consume?

A 1979 paper in *Harvard Business Review* by Paul Farris and David J. Reibstein found a correlation between high prices of goods, high advertising costs, and higher profits. At the same time, companies with lower prices and higher advertising budgets received lower profits from their efforts.[1] By no means the last research on this topic, in 2017, another study found a "positive, significant, linear relationship among advertising expenditure, sales revenue and net profit [*sic*]."[2] These reports, however, address a range of industries in the modern era and do not necessarily reflect realities a century ago.

In examining the late nineteenth and early twentieth centuries in the Rocky Mountain West, we face a number of obstacles to measuring the success of ads and profits. For one, there are few records left that correlate drinkers to brewers—a version of Yelp or UnTapped did not exist. As such, motivation for drinking certain beers over others is largely lost to modern interpretation. Another obstacle is the records from many breweries of their advertising money spent and income received. Many breweries operated on a small scale for short periods of time and therefore either did not keep or otherwise did not preserve these types of records.

Beer terroir advertising cannot be fully measured in a direct correlation to beer sales. It does not seem that brewer's calls for beer terroir were necessarily heard by locals. One could assume that brewers who stayed in business longer were more able to translate their formations of terroir to their customers. However, without direct connections, there are many other reasons—such as economies of scale, stable populations, and capital available—to support longevity besides consumer purchase of beer terroir. Much of the growth of beer produced and consumed may have more to do with the population growth rather than any significant figures. Also, the reduction in consumption on a large scale reflects national trends and does not suggest that drinkers in the Rockies were more devoted or committed to supporting their local breweries in the face of Prohibition than the rest of the nation.

Marketing campaigns did all they could to attract drinkers. Breweries harped on their location in a contest with other options for consumers. But how successful were these efforts? It is difficult to discern in any age how effective marketing and advertising directly connect to sales volume, and in the nineteenth-century Rocky Mountain West, we have fewer records reflecting beverage choices. There are three methods to ascertain whether brewery advertisements successfully attracted local drinkers: consumers' records, production and consumption of beer, and breweries' longevity.

The first of these is difficult, as many drinkers did not record their evaluations of breweries. We do not have an UnTapped app or Yelp to let us know the customers' side of the equation. At the same time, populations in these states were largely transient and so moved across the region in search of extractive wealth. Some of the major cities would have a stable population, but in the mining towns across the region, no matter how good one's beer was, if there was no one in town to drink it, the brewery would not survive.

Another approach to estimating beer terroir advertising effectiveness is to measure overall beer consumption. Arguably, higher consumption would mean that at some level, brewers were successful in convincing the general populace that their beer was indeed worthy. Yet here there are several pitfalls that prevent direct connections. Although this analysis ends when national prohibition begins, states and municipal districts began local prohibition earlier. Effectively, this reduces overall amounts of beer drunk unrelated to how convincing beer terroir campaigns were. Nationally, by 1916, beer consumption dropped forty-five million gallons from the previous year—some of this decrease may be attributed to local and state prohibition.[3]

This reduction coincides with trends in the years leading up to the Twenty-First Amendment. In 1913, nationally, there were eighteen thousand fewer saloons, and in Chicago—a significant brewing capital—imbibers downed over two hundred thousand fewer barrels of beer than the year before.[4] Shrinking opportunity nationally resulted in understandably lower consumption. In the main, less beer brewed and fewer drinking establishments reflected national trends and state political climates.

The reduction in establishments in the 1910s leading up to national prohibition was not an anomaly, however. As the American population grew between the 1880s to 1900s, the number of breweries dropped. Much of this can be explained by the expansion and monopolization of the beer industry by fewer key players. For example, the available capital reported by the over two thousand breweries in 1880 was a combined 91 million compared to the 415 million represented by five hundred fewer breweries in 1900.[5] Fewer breweries with much larger output began to dominate national production.

In the West, brewery closures matched national trends. Pre-Prohibition western states closed up saloons and breweries in huge numbers. Colorado joined the dry states in 1916 and closed hundreds of alcohol-related industries.[6] However, closures reflect state political leanings toward the ills of alcohol. It does not answer whether the breweries' exhortations to their beer's terroir convinced drinkers to consume their products. One could argue that state prohibition meant they were unsuccessful. Yet the week before going into effect, estimates range from two to three million dollars'

Table 5: Breweries per capita, 1880-1900

Year	Number of breweries	People per brewery
1880	2,191	22,892
1890	1,248	50,178
1900	1,509	50,079

worth of liquor that was purchased in the state alone.[7] Clearly, a large portion of the population was not on board with where their state was heading.

These closures do not mean that brewing did not continue apace until local politics ended their run. Salt Lake City reported brewing production of two gallons of beer per person in 1905, adding, "We don't know who got away with our two gallons, but hope it did him good."[8] Other states encouraged beer production brewed with local products—such as the mayor of Laramie in 1908 entreating beer production at "the Laramie Brewery from barley grown on the Laramie plain."[9] If consumers had not bought into the beer terroir of Laramie Brewery, it seems the mayor had.

The reduction in breweries and other considerations above do not seem to correlate to the tastiness of beer terroir produced in the Rockies or elsewhere. Explanations are more in line with the consolidation of the brewing industry and the proliferation of dry towns and dry states. Despite these economic trends, there continued to be a desire for alcohol and for beer.

Perhaps the clearest sign of the acceptance of beer terroir among drinkers comes after each state voted for local prohibition. If the beer was good only because it was intoxicating, then we could assume to hear no more about it until after prohibition was lifted. The existence of "near beer" is a testament against this. By definition, near beer was not intoxicating, and yet there were wide-ranging calls for it among the breweries that still operated. Herman Hamres of Utah presented at a conference a list of beers allowed under the new law that "have the characteristic tang and flavor of beer."[10] Of course, authorities pushed back against near beer and argued that it was in fact intoxicating, and licenses to sell it were denied routinely.

Although some near beer certainly was intoxicating, that likely did not hold true for every iteration. Brewers of near beer continued to produce their goods and people continued to drink them. In Idaho, one "near beer tasted so much like the beverage that made Milwaukee famous that they had to knock on wood to remember they were in dry towns."[11] In Denver, there remained five major breweries by 1915—many smaller operations had been gobbled up by the larger ones. Of these, three opted to brew near beer "that looks like real beer, tastes like real beer," with the exclusion of alcohol the only notable, unfortunate difference.[12] The other two breweries closed up shop.

If we include near beer, then there appears to be support for continued appetite for Rocky Mountain beer among consumers. Brewery numbers dropped but overall production increased over the decades from the Gold Rush to Prohibition. Near beer slaked the thirst for locally produced beer. This overall trend accounts for some of the best evidence to support that people bought into beer terroir produced by local brewers. If not, then we would have seen precipitous drops in beer produced regionally supplanted by imported beer from other states. In addition, the staying power into the near beer era attests to people's appreciation for what their local brewery produced and the ongoing desire for more of that rather than imported near beer.

The last element to examine the appreciation of local beer terroir among drinkers is the longevity of breweries. If they, in general, last over the years with significant investment, then people must be buying—and drinking—the beer terroir they were selling. Of course, breweries, like many restaurants, experience crushingly high failure rates as businesses. This does not diminish the fact that some of these breweries conducted significant investments. A promise of the future led to investment—sometimes not supported by a commensurate increase in revenue.

Brewery failures could be due to failure to produce a quality product, shifting populations, or a reflection of industry trends. One of the major causes of closures across the board is the regional movement to prohibition. People moved away from breweries and saloons, and in Idaho in 1914, "three of the largest breweries closed down during the last year because of insufficient patronage."[13] Whether this was a shift away from locally produced beer toward large-scale mass-produced beer factories importing their products or a rejection of the beer produced within Idaho is unclear. It could likely be a multiplicity of causes.

In the 2020s, Budweiser sent a transgender woman beer cans with their face on them to celebrate a year after transitioning. The cans were sent only to Dylan Mulvaney, not distributed widely. This outraged many conservative consumers who saw this as a rejection of their values and the values they had ascribed to Budweiser's products. Today, as in the past, beer holds many overlapping meanings. In this instance, American consumers imprinted politically and culturally conservative views onto cheap, light beer. Here, the main labels were Budweiser and Bud Light.

These consumers took to social media to publicize their indignation. Promising a boycott, Budweiser's stock plummeted. The outraged saw this as a victory and a testament to their successful anger. Yet it is highly unlikely that a few enraged consumers of cheap beer were sufficient to tank one of the world's largest international corporations. Instead, this dip was in response to much longer-term—and larger—forces pushing down the value of Budweiser; the outrage was simply the catalyst. Despite an advertising blitz—on top of their already massive advertising budget—Budweiser was unable to revert this trend.

The answer, according to Vikas Mittal, a Rice University professor of marketing, lies in the quality of beer. Simply put, "Advertising cannot compensate for mediocre or sub-par quality."[14] We can apply this to brewers and consumers in the Rockies at the turn of the century. Brewers advertised beer terroir as widely and effectively as possible. Frustratingly, there are simply too few sources to understand how well beer terroir touted by breweries translated to the palettes and pocketbooks of drinkers.

Brewers advertised their beer terroir and collectively built a language of beer in the Rockies. Yet while they were doing this, they were combating a number of national trends that worked against them. This makes it difficult to recognize how well consumers bought into beer terroir. Based on the survival of records of beer consumers, there is little to show if drinkers drank beer because they were convinced of beer terroir, because it was cheap and available, or for any of the plurality of other reasons people choose where or what they drink. The reduction in breweries could also have a number of causes that do not relate to how well drinkers connected to the terroir of their beer. Prohibition, mobile populations, the extreme difficulty of the brewing industry, and other challenges outlined in this study were all equally likely to contribute to the failure of a brewery.

9

Exclusions

In 2015, Oregon Rogue Beer, a major brewery in the Pacific Northwest, created a beer using yeast combed from their brewmaster's beard.[1] John Maier's beard had accumulated yeast microbes in sufficient numbers to collect, ferment, and produce a beer known as American wild ale. Done as a fundraising stunt, Beard Beer's proceedings went to support—fittingly—No Shave November. Unsurprisingly, this beer generated a great deal of public comment ranging from appreciation of the craft to outright disgust.

The choice of style belies its message about the permeable relationship between brewery workers and beer. What sets American wild ale apart from other beer styles is the enhanced role microbes play in relation to traditional yeasts.[2] Within this category are Brett beers, mixed-fermentation sour beer, wild specialty beer, and straight sour beer. In these, the influence of the environment of brewing plays an outsized role.

Brewery laborers and their product are often not discrete entities but a continuum of organic material. Wild ales are a particular style brewed with local yeast. Meaning that the yeast that hangs in the air or clings to laborers' clothes are what fall into the open vats for fermentation. Workers bring in yeast on their clothes and bodies, which modern brewers prohibit as much as possible from infecting the brewing yeast. As Maier's actions proved, the barrier between humans, the environment, and brewing is more malleable than we may like to think.

Beer terroir is an inclusive, near-universal term embodying the separate aspects of brewing. However, in the nineteenth-century Rocky Mountain West, there were notable exclusions. Brewers, marketers, and beer factory owners did not include their workers or equipment in public arguments for their beers' superiority. This telling admission in their construction of beer terroir is a reflection of their attitudes toward their laborers and capital expenses in an era of big business and unrestricted capitalism.

Labor

Laborers within Rocky Mountain breweries struggled with management in a series of labor disputes, strikes, and unionization efforts that antagonized the relationship between brewery owner and worker. The terroir of beer includes brewery laborers, and the ways organic and biochemical masses interact within breweries undoubtedly contribute to beer terroir. People moving in and out of these spaces carried with them microbes, yeast, bacteria, and other microscopic elements that must have found their way into batches of beer. Despite this scientific grounding, brewery owners were loath to attribute the quality of their product to laborers in an effort to reduce workers' power in an era of labor disputes. Without laborers there would be no beer; nevertheless, laborers were excluded from a key component of beer terroir.

This is not to say that all brewery owners looked disparagingly at their workers, and in many cases, they could ill afford to. Major breweries like Coors were not commonplace in the late nineteenth-century or early twentieth-century Rockies. Much more common were small-scale businesses that provided a modest amount of beer to their community. There certainly were labor violations and difficult working conditions. Brewing, after all, involves heavy equipment, boiling water, and a high chance of injury.

Unionization efforts proliferated in the late nineteenth and early twentieth centuries. In the Rocky Mountain West, many labor disputes concerned precious mineral extraction. Strikes and efforts to break them led to violence across the Rockies, and breweries were not immune to collective bargaining efforts. Typically represented by the Brewery Workers Union, brewery workers petitioned for better working conditions within the Rockies.

News of strikes—causes and successes—reached brewery owners and workers in several forms. Some of the more prevalent in the Rocky Mountain West prior to Prohibition were the *Denver Labor Bulletin* and the *United Labor Bulletin*. Both reporting out of Denver, their bylines largely centered on labor issues at Coors in the 1910s. One of the largest and thus most influential brewers in the region, how labor disputes arose and were settled in Golden, Colorado, informed the whole industry. Labor success or failure informed both workers and owners in other Rocky Mountain towns how far they could push the other and what kind of industry-standard relationship there was.

Reports on strikes or union efforts in other states gave alternating hope to brewery owners or workers, depending on their outcomes. One issue in the Rockies that severely threatened both was the rise of prohibition. Maneuvering beer around temperance laws had been politically and legally messy as states voted on statewide prohibition. *The Salt Lake Tribune* reported on the end of a brewery strike in Montana in 1909 that came to an equitable resolution, but stated that "the greatest brewery strike, and one that has no end, is that of the prohibition club."[3] As national prohibition and enforcement of a national ban on the production, distribution, and consumption of alcohol grew, workers and owners clashed over an uncertain economic future.

One such clash occurred in 1916 Cincinnati. In a statewide vote in Ohio on the question of prohibition, the wets won a forty-thousand-vote victory.[4] As a result of their success, brewery workers demanded a wage increase. Voting down prohibition meant the continuation of brewing, and thus brewery workers, for their part in the vote, argued they should be rewarded. After all, laborers "have been among the leaders in the fight against prohibition," and brewery owners ought to recognize that support for their continued profits.[5] News of strikes in major beer-producing cities such as St. Louis and Milwaukee circulated in newspapers and magazines in the Rockies, keeping people abreast of demands and results. The well-reported victory of an early 1900 strike in St. Louis that ended with "an increase of wages and shorter hours" must have had an impact on brewery workers' hopes and brewery owners' nightmares.[6]

In the Rocky Mountain region, unionization and labor movements commonly include mining and industrial manufacturing. However, in many states, brewery workers were counted among the number of card-carrying union members. In Montana, Mr. Donahue—the president of the Montana State Federation of Labor—met with striking brewers in the state in fall 1910 and discussed next steps toward a favorable resolution.[7] The brewery strike in question at Lewistown, Montana, ended in February the next year.[8]

Though brewery strikes were not frequent in the pre-Prohibition Rockies, general unionization and labor disputes were. Brewery workers operated within an environment of healthy, robust, and active unions. Some brewery workers joined different unions, such as Brewery Workers United or other state-based organizations. In all, brewery owners saw a vibrant labor movement in the Rockies and, in terms of terroir, would not have wanted to give credit to their superior beers to the workers who created them. Providing laborers with credit for beer terroir would have been tantamount to capitulating in any current or future labor dispute. As such, brewers excluded laborers from their depictions of their beer terroir and what made their beers special. Location, not labor, was given credit for quality lagers.

Capital Purchases

In much the same way as without workers, there would be no beer, brewers were also reluctant to attribute the quality of the brews to their equipment. Yet equipment and brewing structures received more attention than workers. It seems brewery owners were more willing to make the connection between quality equipment and quality beer than they were to connect quality workers to their beer equivalent. Equipment was a safer, less risky aspect to include in beer terroir than workers. After all, kettles and grain elevators may break, but they do not go on strike.

Many newspapers and ads praised modern breweries in their locale. Some argued that good beer could not be made without modern equipment, though they often failed to specify what that meant. Reflecting attitudes within the Rocky Mountain West, a reporter in Keota, Colorado, responded to the growth of statewide support for prohibition. They argued that good beer needed to be produced at a larger scale, stating that "to make a good, wholesome and palatable article requires the best technical appliances of the modern brewery."[3] In essence, good beer was the answer to prohibition, and good beer required modern equipment. Breweries that made additions, renovations, or began fresh were celebrated for using modern technology, processes, and equipment.

Funding for breweries came from initial investment or, for existing breweries, through mergers and acquisitions. Increased capital through various sales and stock options allowed breweries to expand and grow. In doing so, a major selling point was the local terroir. In 1906, Havre Brewing

Company in Havre, Montana, went through a reorganization that aimed to increase capital stock by $15,000.[10] Town boosters in support of this business venture understood what it meant for the economic health of their town. Drawing on beer terroir, one townsperson noted, "Our water was pure as pure could be and well adapted to the manufacture of beer."[11] Additions to breweries came at an expense, and investments were grounded in marketing beer terroir.

New establishments of breweries were often events of note. Brewery owners in the Rockies wanted to ensure the future of their business ventures. Advertisements in the local paper could be more or less informative for consumers. A Carson City paper unimaginatively noted, "The new brewery is a-brewing something."[12] A more expedient way to appeal to consumers was by both connecting their brewery to the growing beer terroir of the region and making note of their modern and sophisticated equipment or facility. As above, water purity often was the main beer terroir category. But modern equipment and facilities often played a crucial role in presenting new breweries to the public. Henry Droppleman, a brewery promoter in Albuquerque, extolled the virtues of a new brewery that was "modern in every detail" and that would "be capable of turning out the best quality of lager beer that can be manufactured."[13]

In general, additional capacity, renovations, or new breweries signaled longevity for mountain communities. Companies willing to invest in brewing meant diversification of their economy and less reliance on mineral prices. The Anaconda Brewery in Montana underwent a merger/acquisition and became Petritz, Steiger & Faul in 1895. With that came renewed capital investment into the facility to the tune of $30,000. The town of Anaconda understandably gushed about this influx of cash and increased brewing capacity. Noting the additions, one denizen proclaimed, "Their enterprise and pluck speaks volumes for their faith in Anaconda as a permanent, prosperous city."[14] As discussed in chapter 5, breweries signaled stability for towns that had been whiplashed by mining and extractive industries' boom-bust cycle.

Brewery owners and operators paid attention to growing labor rights movements in the American West and across the beer industry. Responding to this, they keenly excluded their workers from their conceptions of beer terroir. No arguments rang out that one's beer was better because of the workers within their walls. They drew a clear line between product and producers. Scientifically, this was not quite accurate. As Rogue Brewery's radical beer shows us, the microbial barriers between humans and brewing are less rigid than we may suppose. Laborers brought into the brewery not only their bodies but also their skills, assumptions, traditions, heritages, and microbial freeloaders. All of these were part of their final product, even if not part of a publicly pronounced beer terroir.

Equipment and modern facilities received more praise, but even still, brewers and boosters were not claiming superior beer because of their equipment in the same way that they did for water, grain, or other ingredients. Fine equipment and modern brewers were essential, but their significance lay more in the capacity to produce more than in the imprint capital purchases laid on the beer's taste and profile.

10

Nostalgia & Memory

Nostalgia and memory are powerful forces in the human psyche. They influence our interpretation of our personal histories as well as our present. Food and taste attach to our memories, and scholars have investigated the ways and degree to which we can taste memory.[1]

Memories reflect neither the exact recollection of what happened nor our own experiences. Rather, they present us with a mental image of what we imagine transpired. This can be closely related to what actually occurred but is never a true reiteration of the past. The same mental tricks our minds play are similar in history beyond our lives. Popular visions of the Old West in movies, novels, and the like may capture a portion of what life was like. Nevertheless, they will always impose—intentionally or not—aspects of falsehoods into the portrayal.

The Rockies are particularly prone to manipulation of memory, and white nostalgia about this region is powerful. Ideas about miners, cowboys, pioneers, Native Americans, and other images are powerful in Americans' real or perceived memory of this place. Promoted in movies, games, pop culture, and elsewhere, myths of the "Old West" are active in the present and create ideas and visions of the past. Craft beer in recent decades affirms the terroir built pre-Prohibition through its pride in local ownership and defenses against Big Beer, promotion of local ingredients, and usage of historical figures and narratives. Although the focus here is on the Rocky Mountain West, breweries across the US employ history in building their identity and expanding their marketing.

This chapter covers a wide time period, from Prohibition until the modern craft brewing boom. It examines the ways breweries connect themselves to the past in an effort to validate, refute, or confirm their brand's place in the beer industry. By combing through history, brewers use myths, memories, and appeals to historical nostalgia in order to market their beers. In essence, through drinking this beer, from this brewery, you are participating in a grand, unbroken history of the Rocky Mountains. While not necessarily true, it is the sense of historical terroir that makes this connection in consumers' minds and bellies possible. Potent feelings of historical connection, despite the various historical breakages, abound.

This chapter leaps several decades after Prohibition to examine how breweries resurrected legends, histories, and regional flair to sell beer amid the craft beer revolution. Ostensibly begun in the 1980s, the craft beer revolution has undergone several iterations and countless upheavals. The point of this discussion below is to see how, amid the numerous breweries popping up or going under, some tied themselves to their terroir to maintain solvency. For example, Colorado Native is a beer brewed with only Colorado ingredients and sold only within the state. They are intentional about this when it is likely the case for many other breweries; they just have not included it in their ethos. Yet of them and the countless other breweries in the Rocky Mountain region, this chapter asks, how does the past sell today?

Post-Prohibition Beer

Prohibition destroyed the American brewing industry. Western states often implemented statewide prohibition prior to the Eighteenth Amendment, but Prohibition sounded the death knell for the remaining brewing industry. Those that had struggled to overcome local edicts wilted under the force of national law. The dry decade eliminated the foundation and lifeblood of Rocky Mountain brewing: It severed trade lines, curtailed hop production, devastated local breweries, and after such a period, pushed a generation of skilled laborers out of brewing. The loss of physical, network, technical, and financial capital meant that only a few breweries could survive. Most often, as in the West, these were monolithic beer factories with the resources to weather the drawn-out drought.

The fate of post-Prohibition beer in the US receives scant attention compared to the more recent phenomenon of the craft beer boom. Yet it lays an important foundation for that boom. Ranjit Dighe wrote a humorously apt subtitle for a chapter that explores this period: "How American beer got to be so bland."[2] Indeed, this was what happened: a blanding of beer as a few breweries consolidated power and made beers cheaply. Reducing the variety of beers offered and using cheaper or less malt led to a thinning of beers available for people in number, options, and taste. The great profusion and variety of beers today likely tower over the pre-Prohibition beer offerings, but this is not to diminish the cast of casks that filled saloons before the 1920s.

Essentially, the lead-up to national prohibition was foreshadowed by numerous state-initiated prohibitions. As such, breweries began to close up due to legal pressure, mounting public opinion against them, or declining consumer bases. This happened across the US but particularly in the American West, which initiated and maintained statewide and local prohibitions before the Eighteenth Amendment and left them in place after its repeal. As we have seen, some breweries saw the direction the US was headed and sought to join, and thus protect beer, rather than fight Prohibition—an effort to split hairs that ultimately failed. This amendment obliterated smaller brewers, and only the largest breweries with significant capital and capacity to diversify survived.

After Prohibition's repeal with the Twenty-First Amendment, it was not simple enough for small brewers to begin fermenting again. The large beer factories that survived were filled with jubilee and began rolling out kegs of beer almost simultaneously with the law's ratification. This immediately put smaller brewers at a disadvantage; large brewers had networks, equipment, and economies of scale to crush upstart brewers. Brewing is a capital-intensive labor to start. The money needed to hire trained employees, buy expensive equipment, and purchase ingredients allows room for little error or experimenting. Further, an entire generation of brewing skills, knowledge, and expertise was lost between 1920 and 1933. These factors left potential new brewers few resources to work with or to pull from.

Finally, even if home brewers possessed the equipment and skill and could appeal to a consumer base, they had to contend with legal barriers. Some states had local laws prohibiting brewing, and even though the national ban was lifted, this did not change local mandates.[3] In addition, to prevent a return to the drunken-filled days temperance sought to destroy, layers of legal restrictions existed on state and federal levels. Some were holdovers from Prohibition, while others were legislated afterward to

retain some control over alcohol production and consumption. These limited distribution and production and placed bars too high for smaller brewers to reach. As a result, the American brewing industry after Prohibition
remained in the hands of a few major national brewers. Until the 1980s
craft beer revolution, which required the reversal of many of these laws,
repeated hammer strikes of laws and the disruption of World War II meant
a quick shrinking of the American brewing industry in the early twentieth
century.

Selling Nostalgia

With this in mind, how then to explain the use of history in constructing
terroir from more recent breweries? How can companies not more than
a few years old, or perhaps only a few decades old, connect themselves
across the great chasm of 1920–80? What drew brewers to the past for
inspiration, and how did history become an intangible part of individual
breweries? That is a significant stretch of time, and it takes a mixture of
imagination, savvy marketing, and tapping into humans' deep, instinctual
connection to nostalgia and memory.

Nostalgia maintains a powerful pull on human imagination. It plays a
role in attracting tourists and is used extensively in advertising.[4] It fosters
a desire, need, and industry for edible tourism, as people travel to certain
places to satiate a culinary curiosity or desire.[5] Advertisers peddle food and
drink experiences tied to certain places. This both fuels terroir and feeds
off established conceptions of terroir. If a particular place has a reputation
for a certain product, then marketers can employ this to sell their product
or sell packaged experiences. Paying for a product or a place increases the
exclusivity of that product and deepens the terroir attached to it. In this
symbiotic relationship, savvy marketers can continue to build a reputation
for a product tied to a place. The more people buy, the more established the
terroir becomes. Essentially, once people begin to believe it is worth paying
a premium for a product, like beer, that comes from or exists only in a certain place, they will justify their decisions. In effect, they become ambassa-
dors for the product, encouraging their peers to participate in the same way
and equally contribute and expand the value of the terroir.

The beer world has a few major examples of this phenomenon. Guinness, famously, is widely believed to taste better in Ireland. There are several

scientific reasons for this, centrally that beer does not travel tremendously well. But many theories abound as to why it tastes better. A likely and persuasive argument is simply that one is in Ireland, and drinking a beer so closely connected to the land is an experience that transcends simple freshness.[6] Few travel to Ireland explicitly to drink Guinness in its homeland, but certainly it is part of many itineraries.

Craft beer has leaned into theories of local production and provenance. A study in 2018 examined how much extra people were willing to pay to consume local beer instead of imported beer, where the definition of local was left to the consumer.[7] On average, people were willing to pay twenty-five to fifty cents more per beer if they perceived the product as local. Studies such as this can help us understand the success of brewers in the nineteenth century who demonized imports and defended local beers.

Breweries across the US have attempted to harness history in an effort to boost sales or otherwise stand out from the crowd. Commonly, craft breweries use this to connect their recent history to a larger narrative of brewing in the US. Craft breweries, although many have been bought and combined with larger entities and thus are no longer "local," do still contain that characteristic. This helps them bridge a temporal gap to boost their credibility and status. In contrast, major American breweries that have indeed stood the test of time and have well over a century of essential continuous operations use their own history to reinvigorate sales.[8] For example, Pabst has attempted to rerelease beers that have not been brewed for decades (or in some cases, a century) and has released beers reminiscent of long-dead recipes.[9] Using history as a weapon to wield against upstart breweries, Pabst hopes to appeal to senses of nostalgia and history in order to sell their beer.

Tasting the Rockies

The manifestation of history in beers and brewing is rampant in the American West. As in many other arenas, the Rocky Mountain West is an ideal place for nostalgia, mythmaking, and storytelling that reflect an idealized and heroic past.[10] Historians have obviously doubted the validity of these memories, but the fact remains that the West has a rich past that marketers and advertisers have long applied to their particular products. Breweries are no exception here.

The Rocky Mountain West has attracted its own share of beer tourists. Ostensibly not just coming for the beer, visiting local breweries and tasting the local brews is a growing attraction for visitors.[11] Breweries employ nostalgia and memory in ways that suit their individual needs in their time and space. What connects them is that whether it be in minuscule, nuanced decor or beer descriptions or in major broadcast messages, the past is very much a part of contemporary brewing.

Brewed with ingredients solely obtained from Colorado and sold only to the same state, Colorado Native Brewery is an offshoot of Coors.[12] With names like Native, Local, High Country, or Western Slope, their beers take after regional nomenclature and history. Directly, this brewery clings to the history-steeped physical geography and brings it into the core of their brand. Not the only group to use regional terms, local history, geographical boundaries, and state pride to sell beer, Colorado Native highlights the continuation of terroir from the pre-Prohibition period to the current era in American brewing.

Others continue the tradition of terroir and highlight the goodness of their environs. Grand Teton Brewing in Idaho boasts, "Teton Valley, Idaho is the best place on earth to craft beer. Our water is glacial run-off, filtered over 300–500 years by Teton Mountain granite and limestone before it surfaces at a spring a half mile from the brewery. Teton Valley grows the world's best malting barley, and Southern Idaho includes some of the finest hop farms in the world."[13]

Drinking History

Marketing through naming is not the direct cause of beer tourists. However, in a sea of breweries and beers, whatever strategy one can employ to capture a larger segment of consumers is a worthy avenue. Mining history and rolling out beers that are reminiscent of what one imagines of the historic Rocky Mountain West is a useful strategy. Naming is one such storytelling strategy that immediately connects a beer to a historic interaction.

In Idaho Springs, Colorado, a historic mining town, a brewery emerged that connects to this past. Tommyknocker is a folklore creature that came to the Rockies via Cornish miners. If supplied with kindness and food, Tommyknockers repaid miners with warnings of imminent wall collapse or other catastrophes or led miners—by responding to tapping on the mine walls—to rich ore veins.[14] Taking the name of a mythological mining entity connects the brewery, a business around since 1994, to a past over a century ago. Neatly jumping the middle decades, this promotes a sense of historical continuity and connection rather than an obstructed past.

History also highlights people and places that have longer histories than the nineteenth-century mining rushes. Headquartered in Albuquerque, New Mexico, Bow & Arrow Brewery is a Native American–run brewery that focuses on connecting their beer to the historical and current Native inhabitants of the Southwest.[15] One of their endeavors is not only to bring attention to the long history of Native Americans in the American West but also to collaborate with other breweries across the US to show whose land they stand on. This effort and their brewery combine various elements of terroir and brewing that force drinkers to reckon with the complex history of brewing.

In many ways, then, the advertising schemes of more recent Rocky Mountain breweries mimic those of their pre-Prohibition forebears. Claims to higher—sometimes literally—quality ingredients and historic connections appeal to consumers to participate in the environment and nostalgia bubbling within each pint of beer. History is an integral element of terroir, and modern brewers are able to take advantage of the labors and efforts of brewers decades prior. The groundwork that brewers in mining towns throughout the Rockies laid became the rich historical veins that recent brewers tap into and follow to extract relevant marketing material and inspiration for beers.

Brewers today are using terroir and its subcomponent of history to validate, legitimize, and authenticate their position in the beer community. In this millennium, as the craft beer movement struggles to find its next breath of life, the role of authenticity is key. Craft brewers who remain independent are dwindling, and like in the post-Prohibition years, big breweries dominate the market and make entry difficult.

11

Fermenting Terroir & Future Directions

This research and writing for this work took place during COVID-19. Place and space assumed a much greater prominence in many people's lives. One's locality—favorite restaurants, frequented coffee shops, regular watering holes—came into hyperfocus. Initially, fears over closures of small businesses, like breweries, gripped people. People came together to support their local businesses, even if that meant buying gift cards there to be redeemed once it was safe to return. Stories abounded of community efforts to protect and preserve local institutions.

Efforts such as this were fertile ground for thinking about how place is central to the way we navigate the world. One frequented chain restaurants to support not the brand but rather the local wait and kitchen staff. What people were close to physically and emotionally mattered greatly.

It is in this environment that it is fitting to reexamine the terroir that is constructed incrementally. Brewers in the nineteenth- and early twentieth-century Rockies could be assured of very little in their future. Their townsfolk and consumer base might move away; their water might be polluted beyond repair; sufficient good-quality grain and hops may not arrive. Yet in their obstacles, day by day they built a reputation around their beers. In competition against one another, they employed similar language and tactics to assure their customers that theirs in fact was the purest. However, the purity they all ascribed to reflected similar climatic, historical, and geographical realities. They brewed with the same "pure mountain water" as it ran down mountainsides.

Over time, terroir and tastes have ossified, solidifying lines between state borders. Rather than thinking of the West as a distinct beer region, one considers Colorado or Utah beers as coming from separate areas. Occasionally, beer regions are even more narrow, such as on a city-by-city level. One can continue to pare down the size and scope under study. What this work hopes to do is help us understand both how regions do not conform neatly to state boundaries and how terroir can be minutely and infinitely subdivided into smaller and smaller areas.

In my home state of Colorado, for example, we are often held up as having great beers. More than that, the Great American Beer Festival is held in Denver, which itself is perceived as a beer center. Further, there are certain neighborhoods known more than others for their beer. And so on.

This work has attempted to survey a number of terroir components in an effort to help us understand the importance of place in the beer we drink. As many as have been touched on here, there are smaller spaces to examine and other factors of terroir that impact brewing strategies, flavor, and consumption patterns. If this helps beer producers and consumers rethink the history of and nonpalatable influences on their beers, then this has been a success. Beer, as is the case with many foods, is a product of history, culture, and values.

Climate and geography are the foundations of brewing. Euro-American brewers chose their ingredients and practices in line with what was available and palatable thousands of years ago. This set the stage for centuries of brewing patterns. When white settler-colonists migrated to the Rockies, they encountered a quite different environment that had been shaped by eons of tectonic movements. This was the canvas upon which brewers built their breweries and constructed a new beer terroir.

History directed the ways in which American brewers operated in the Rocky Mountain West. In conjunction with climate, this birthed beer terroir in the Rockies. Producers and consumers wanted a particular type of beer with specific ingredients and taste profiles. The beer styles brewed—mostly lager—were based on the heritage of Germanic brewing. This history came into collision with the climate of the region and set up numerous problems for aspiring brewers. It is here that beer terroir emerges most prominently. The need for brewers to reconcile their new home with their old beers meant that they needed to employ history to bridge the gap. In constructing beer terroir, they navigated the expectations and perceptions of their consumers.

Water, in great quantities, is instrumental in brewing beer. Rocky Mountain brewers found available, dependable, and clean water in short supply. In their pursuit of water, they walked a fine rhetorical line when they gave greater definition to budding beer terroir. They argued—with much basis in fact—that water sources were unclean and carried diseases. These were often due to the very enterprises that brought people into the Rockies in the first place, such as mining or lumbering. In the same breath, they emphasized the cleanliness of the water used in their particular brews. The mental gymnastics required to distinguish the purity of water, most likely from the same source, contributed greatly to taste perceptions for consumers.

Grain as malt provides the backbone of the beer and terroir. Brewers were able to access grain as local farmers expanded their fields and put more acreage under the plow. In this, farmers and brewers worked together to explain to consumers why their barley was superior to others. Although alternate grains were used and adjuncts were incorporated to extend the malt supply, the use of barley has a deep history and importance in Euro-American brewing. The best brews used the best barley—a statement that brewers were sure customers were aware of.

Hops painfully reminded brewers that their region had limitations to its resources. They financially and logistically maneuvered in their economic and physical space to ensure an adequate supply of brewers' gold. Here we understand the borders of terroir and how negatives may in fact support the beer terroir. Meaning that the struggle to acquire hops is part and parcel of beer terroir in the Rocky Mountains. It is not only the benefits that contribute to terroir, but also the burdens.

Local production of beers—loosely defined—expands our framework of beer and how people in the past considered their foodways. What people consider beer in a certain time and place is flexible. Home industries produce a variety of fermented beverages to provide nutrients, calories, and a flavorful, safe beverage. Spruce beer was preeminent among these productions, and its usage ties into a long-standing American history of brewing.

History and nostalgia play a major role in cementing the beer terroir of a region. Brewers continue to produce beer in the same cities as they did a century ago. History has become an essence of the flavoring and ambiance of drinking certain beers—not only in cases where the beer has been produced continually for decades but also using historical stories, narratives, and place-making in advertising and beer naming.

One of the most attractive, compelling, and frustrating qualities of terroir is its endless avenues for examination. There is no end to the ways in which we can subdivide, identify, and further explore why foodways taste different. Our perceptions of taste are easily manipulated and informed. Further study of beer terroir, I believe, will enhance our understanding of the intersections of beer, agriculture, geography, and flavor. In all, this work aims to lay a foundation. A framework and approach that other scholars can take and apply to other sectors of the world. Every region has a beer terroir, and exploring its routes, avenues, and winding roads will hopefully illuminate aspects of how humans perceive their world of taste. Further research of beer terroir—for example, in the American South—and how it differs from other regions in the US would add depth to our understanding of how distinct regions of the US developed their own beer culture. Why and how regions create their own beer terroir or borrow from others contributes to understanding the regional drinking patterns of America past and present.

One popular appellation for beer-producing cities is the "Napa Valley of Beer." In Colorado, this title has been given to Fort Collins and Denver while also applied to various other cities throughout the US.[1] Napa Valley connotes the perfect environment for certain products, most obviously wine in the US. It represents an idyllic situation for growth and production. The fact that in reading Napa Valley, one begins to conjure up images of vineyards and grapes, even if one has never been there, speaks to the power of terroir and imagination discussed in this work. Further, how breweries and promoters have appropriated this terroir-laden term and applied it to beer cities attests to the fundamental and unspoken power of terroir and provenance.

Their ideas about terroir and brewing also encapsulate race and gender, which are not discussed in detail here. Women performed roles in breweries that conformed or came into conflict with standard views of gender. Likewise, race and ethnicity are crucial to understanding beer production and choices. Some focus has been on the ethnicity of brewers and drinkers, but much more could be done to understand the people who were included and excluded or recoiled from the beer industry.

Beyond beer terroir of different regions of the US, it would be productive to investigate the ways in which other alcohols have borrowed concepts of terroir historically and in the modern Rocky Mountain region. Beer terroir is one such alcohol that has co-opted ideas of terroir. My argument here is that the history of selling beer via terroir is an older, established tradition. Nevertheless, terroir as a selling point has grown and expanded in the Rockies. Colorado wineries, distilleries, and other alcohols have recently attributed their superiority based on their location. Imbibing histories of nostalgia, 10th Mountain Whiskey connects their distillery to the 10th Mountain Division that trained and operated out of Camp Hale during World War II.[2]

Beer terroir built by brewers in the Rocky Mountain West between the Gold Rushes and Prohibition was constructed piecemeal and over time. It was not a conscious goal but rather something produced step-by-step as brewers independently tried to ensure their short-term success. They faced challenges in production and marketing and, at almost every turn, manipulated these obstacles into benefits and unique aspects that improved their beer. Understanding beer terroir across a wide region amplifies our understanding of the congruences and continuity of the Rockies. Geographical and environmental influences of beer terroir set this region as a distinct space against other spaces within the US, with ramifications that continue into beer and brewing culture today.

Notes

Preface

1 Bodnár et al., "Alcohol and Placebo."
2 Bodnár et al., "Alcohol and Placebo."
3 Smeets and de Graaf, "Brain Responses."
4 Linderberger, "Despite the Pandemic."
5 Broderick and Mueller, "Theoretical and Empirical Exegesis."

Introduction

1 Van Horne, *Johanna Livingston to Robert Livingston re: Receipt of Spoiled Beer*.
2 Van Horne, *Johanna Livingston to Robert Livingston re: Request for Livingston to Send Beer and Biskett*.
3 *Account of [[Beer]] and Rum*; Livingston, *Account of Provisions and Beer*.
4 Briant S. Young, "One of Ogden's Many Industries," *Truth: The Western Weekly*, April 4, 1908, p. 22, image 22.
5 Young, "One of Ogden's Many Industries."
6 Young, "One of Ogden's Many Industries."
7 "Local Layout," *Daily Enterprise*, January 21, 1884, image 3.
8 "The City," *Las Vegas Daily Gazette*, March 28, 1884, image 4.
9 "All Sorts," *Morning Appeal*, April 28, 1887, image 3.
10 Hoalst-Pullen and Patterson, *Geography of Beer*.
11 *La Junta Tribune*, March 26, 1898, image 6.
12 Alberts, "Brewed to Stay"; Hoverson, *Land of Amber Waters*; Hoverson, *Drink That Made Wisconsin Famous*.
13 See, for example, Knoedelseder, *Bitter Brew*; Baum, *Citizen Coors*.
14 Foda, *Egypt's Beer*.

15 These are some of the more significant and well-known works or famous and integral breweries in the craft beer revolution. Of course, there are more examinations of major and minor breweries. Harry, *Stevens Point Brewing Company*.

16 Smith, *Fifteen Turning Points*, 2–4.

17 Bennett, *Ale, Beer, and Brewsters*.

18 Briggs, *Brewing*, 52.

19 Briggs, *Brewing*, 52.

20 Briggs, *Brewing*, 1–3.

21 "Lager Beer," *Helena Weekly Herald*, August 7, 1879, p. 7, image 7.

22 *Silver State*, August 7, 1883, image 2.

23 "Our Beer Record," *Walker Lake Bulletin*, January 8, 1896, image 1.

24 Libkind et al., "Microbe Domestication."

25 Farber and Barth, *Mastering Brewing Science*, 4.

26 De Keersmaecker, "Mystery of Lambic Beer."

27 "Yeast Biology," in Briggs et al., *Brewing*, 363–400, https://doi.org/10.1533/9781855739062.363.

28 Appel, "Artificial Refrigeration."

29 "The Tower Brewery Lay-Out," in Briggs et al., *Brewing*, https://app.knovel.com/hotlink/pdf/id:kt003RWZ96/brewing-science-practice/tower-brewery-lay-out.

30 Neihart, "Beer Wars."

31 Young, "One of Ogden's Many Industries."

32 Brewing manuals are the best source of information for this. See, for example, Lars Marius Garshol, *Historical Brewing Techniques: The Lost Art of Farmhouse Brewing*, Brewers Publications, 2020.

1. Terroir & Taste

1 Zhou et al., "Wine Terroir."

2 Zhou et al., "Wine Terroir."

3 Howland and Dutton, "Making New Worlds," 6–7.

4 Swinburn, "Deep Terroir as Utopia," 222.

5 Swinburn, "Deep Terroir as Utopia," 221–22.

6 For more on terroir and wine identity, see, for example, Harvey, White, and Frost, *Wine and Identity*; Gabel, "Wine Origin Authentication"; and Tomasi, Gaiotti, and Jones, *Power of the Terroir*.

7   Dougherty, *Geography of Wine*.

8   Unwin, "*Terroir*."

9   Unwin, "*Terroir*."

10 Harding, "Inventing Tradition and Terroir."

11 Melewar and Skinner, "Territorial Brand Management."

12 Melewar and Skinner, "Territorial Brand Management."

13 For some works on hops and terroir explicitly and by extension, see Van Holle et al., "Brewing Value of Amarillo Hops"; Van Holle et al., "Relevance of Hop Terroir"; and Kopp, *Hoptopia*.

14 Neihart, "Frontier Beer."

15 *Boulder County News*, April 20, 1870, p. 1, col. 3.

16 Kopp, *Hoptopia*.

17 Kopp, *Hoptopia*.

2. Climate & Geography

1   There are a tremendous number of works on the violent, genocidal invasion of white settlers into the Rocky Mountain region and the displacement and devastation of Indigenous communities. Some powerful works on this include West, *Contested Plains*; and Hämäläinen, *Comanche Empire*.

2   Limerick, *Something in the Soil*, 23–26.

3   For more on the social, human, and ideological conceptions of mountains, see Price et al., *Mountain Geography*.

4   Parry, "Geology and Utah's Mineral Treasures."

5   *Encyclopaedia Britannica*, "Laramide Orogeny."

6   US Census Bureau, "Population of Nevada by Counties and Minor Civil Divisions," 1901, 1; US Census Bureau, "Population of Arizona by Counties and Minor Civil Divisions," 1900, 1.

7   Pomranz, "125-Year-Old Olympia Beer."

8   *Lewiston Evening Teller*, August 4, 1906.

9   National Park Service, "Weather."

10 Western Regional Climate Center, "Average Statewide Precipitation."

3. Pure Mountain Water

1   Barth, *Chemistry of Beer*, 69.
2   Native Americans, notably the Comanche, controlled rivers and returned to them seasonally to reaffirm political and economic relationships. For example, see Hämäläinen, *Comanche Empire*; West, *Contested Plains*.
3   Bramwell, *Wilderburbs*, 78.
4   Gatrell, Nemeth, and Yeager, "Sweetwater."
5   See, for example, Hoalst-Pullen and Patterson, *Geography of Beer*.
6   This was common across the US in the nineteenth century until refrigeration became far more accessible and capable. For example, Coors had massive ice production during the winter months. By 1875, only a few years after opening, they had three thousand tons of ice blocks prepared to cut over winter. "Coors and Schueler Brewery Report," *Rocky Mountain News*, February 7, 1875, p. 2.
7   Summitt, *Contested Waters*.
8   Summitt, *Contested Waters*.
9   "Wagener's Imperial Beer," *Goodwin's Weekly*, March 7, 1914, p. 8, image 8.
10 "Wagener's Imperial Beer."
11 West, *Saloon*.
12 "Fear a Water Shortage," *Morning Appeal* (volume), March 16, 1898, image 1.
13 Limerick, Hanson, and Brown, *Ditch in Time*.
14 Bramwell, *Wilderburbs*, 78–79.
15 "The Water Question," *Goodwin's Weekly*, August 26, 1905, p. 2, image 2.
16 Bramwell, *Wilderburbs*, 78.
17 "Water Question."
18 "Water Question."
19 Rush, *Enquiry*, 243.
20 *Delta Independent*, October 24, 1894, p. 1.
21 "Official Nominations," *Livingston Enterprise*, October 29, 1892, p. 6, image 6.
22 G. B. Morse and Jas. Kirkpatrick, "W.C.T.U. Department," *Dillon Tribune*, August 30, 1884, p. 1, image 5.
23 C. S. Burnett, "Temperance," *Salt Lake Herald*, March 11, 1888, p. 14, image 15.

24 *Helena Weekly Herald*, July 3, 1884, p. 4, image 4.

25 *Neihart Herald*, November 27, 1897, image 3.

26 *Greeley Tribune*, March 29, 1906; *Anaconda Standard*, July 30, 1895, p. 3, image 3.

27 Teetotalers were a significant, powerful force in the late nineteenth and early twentieth centuries. Prohibition, the Eighteenth Amendment, is their crowning achievement. See Rorabaugh, *Alcoholic Republic*.

28 See West, *Saloon*; Noel, *City and the Saloon*.

29 "A Prescription of Health," *Daily Independent*, July 10, 1899, image 3.

30 "Prescription of Health."

31 "The North Park W. C. T. U.: Beer: High Grade and Low Grade," *New Era*, June 23, 1910, image 1.

32 "North Park W. C. T. U."

33 "North Park W. C. T. U."

34 Alworth, *Beer Bible*, 58.

35 Rodrigo et al., "Influence of Style"; Alworth, *Beer Bible*.

36 Eumann and Schaeberle, "Water," 97–111.

37 Yool and Comrie, "Taste of Place." This is not the only beer style to be heavily influenced by its place of origin's water source.

38 Smith, *Beer in America*.

39 Flavin et al., "Interdisciplinary Approach."

40 Philpott, *Vacationland*.

41 Duane Smith is a preeminent historian of mining in the Rockies, particularly Colorado. Some of his works include John L. Ninnemann and Duane A. Smith, *San Juan Bonanza: Western Colorado's Mining Legacy* (Albuquerque: University of New Mexico Press, 2006); Duane A. Smith, *Rocky Mountain Boom Town: A History of Durango, Colorado* (University Press of Colorado: Niwot. 1992); Duane A. Smith, *Mining America: The Industry and the Environment, 1800–1980* (University Press of Colorado: Niwot, 1994).

42 For the impacts particularly of silver mining, see Gomez, *Silver Veins, Dusty Lungs*; McNeill and Vrtis, *Mining North America*; and Robins, *Mercury, Mining, and Empire*. For more on the impacts of mining on water and landscapes in the American West, see MacDonnell and University of Colorado Boulder, *From Reclamation to Sustainability*; and Gammons, Metesh, and Duaime, "Overview."

43 "Purifying Water for Brewery Use," *Western Brewer and Journal of the Barley, Malt and Hop Trades*, February 1906, p. 66, https://babel .hathitrust.org/cgi/pt?id=chi.103524575&view=1up&seq=76&q1 =yeast%20sale.

44 "Beer That Will Make Dry States Wet," *Eureka Sentinel*, June 10, 1916, image 4.

45 "Round About Town," *Santa Fe Daily New Mexican*, September 23, 1892, image 4.

46 Philpott, *Vacationland*; Childers, *Colorado Powderkeg*.

47 Philpott, *Vacationland*; Childers, *Colorado Powderkeg*.

48 "Why Beer Is Healthy in Summer," *Salt Lake Herald-Republican*, August 7, 1910, p. 6, image 22.

49 *Grangeville Globe*, October 2, 1913, image 6.

50 *White Pine News*, July 7, 1888, image 3.

51 *Tonopah Bonanza*, August 3, 1901, image 4.

52 "He Condemns Beer but Uses Tea and Coffee Freely," *Ogden Standard*, August 21, 1909, p. 7, image 7.

53 Rorabaugh, *Alcoholic Republic*, 1979.

54 "Temperance," *Evening Standard*, November 1, 1910, p. 5, image 4.

55 "Temperance," *Evening Standard*.

56 "Temperance," *Evening Standard*.

4. History & Culture

1 Bat-Oyun et al., "Who Is Making 'Airag.'"

2 Among others, see, for example, Cutler and Cárdenas, "Chicha"; Logan, Hastorf, and Pearsall, "'Let's Drink Together'"; Mayer, Sayre, and Jennings, "Coming Together to Toast"; Parker and McCool, "Indices of Household Maize"; Moore, "Pre-Hispanic Beer in Coastal Peru."

3 "Selling Beer to Indians," *Salt Lake Herald*, February 6, 1892, p. 2, image 2; "Their Big Mistake: Town Goers Who Wanted to Treat an Indian Kindly," *Anaconda Standard*, February 11, 1897, p. 7, image 7.

4 *DeLamar Nugget*, December 14, 1891, image 4.

5 "Bock Beer," *Morning Appeal*, May 2, 1886, image 3.

6 "Local Briefs," *Daily Inter Mountain*, June 22, 1899, p. 5, image 5.

7 Nelson, *Barbarian's Beverage*, 4.

8 Nelson, *Barbarian's Beverage*, 10.

9 Nelson, *Barbarian's Beverage*, 10; Cabras and Higgins, introduction to *History of the Beer and Brewing Industry*, 3.

10 Some posit that brewing emerged independently in other regions, such as what is now modern China. While entirely possible, I believe that brewing spread along trade routes and migration patterns. Beer is a flexible, enjoyable, slightly intoxicating source of nutrition and would have been a boon to agricultural societies. See McGovern, *Ancient Brews*, 57; Cabras and Higgins, introduction to *History of the Beer and Brewing Industry*, 3.

11 Nelson, *Barbarian's Beverage*, 83.

12 Nelson, *Barbarian's Beverage*, 110. Richard Unger goes into detail in his work *Beer in the Middle Ages and the Renaissance*.

13 Nelson, *Barbarian's Beverage*, 107.

14 Bennett, *Ale, Beer, and Brewsters*, 80.

15 See also Unger, *Beer in the Middle Ages*; Unger, *History of Brewing in Holland*.

16 Alworth, *Beer Bible*, 17–20.

17 See Hogan and Brady, *Great Chicago Beer Riot*.

18 Statistica, "Beer Production Worldwide from 1998 to 2020," accessed April 5, 2022, https://www.statista.com/statistics/270275/worldwide-beer-production/.

19 Cabras and Higgins, introduction to *History of the Beer and Brewing Industry*, 5.

20 Knoedelseder, *Bitter Brew*, 15; Ogle, *Ambitious Brew*, 9.

21 Ogle, *Ambitious Brew*, 12.

22 Ogle, *Ambitious Brew*, 12.

23 Alworth, *Beer Bible*, 400.

24 Noel, *City and the Saloon*, 20.

25 Andrews, *Coyote Valley*, 27.

26 Smith, *Trail of Gold and Silver*, 13–16.

27 West, *Saloon*.

28 "Great Falls Notes," *Neihart Herald*, May 20, 1893, image 1.

29 West, *Saloon*.

30 Lears, *Rebirth of a Nation*, 123–26, 139.

31 For a compelling history of the Plains Native Americans that resolutely protected their territory and only over a long period of time

lost their ground, see Hämäläinen, *Comanche Empire*. See also Fiege, *Republic of Nature*, 44–50, 254; West, *Contested Plains*, 117.

32   West, *Contested Plains*; Fiege, *Republic of Nature*, 145.

33   US Census Bureau, "State or Territory of Birth (Continued), Country of Birth, Foreign Parentage, Persons of School Age, Males of Militia and Voting Age, Conjugal Condition, Dwellings and Families, Indian Population, Alaska, Form of Schedule and Method of Tabulation," in *Eleventh Census*, vol. 1, pt. 1 and pt. 2, *Report on Population of the United States* (1895), clxvi.

34   Lears, *Rebirth of a Nation*, 159. For a great text detailing Prohibition—its causes, faux pas, and otherwise—see Rorabaugh, *Alcoholic Republic*.

35   Etheredge, "Prohibition at 100"; Ogle, *Ambitious Brew*, 132.

36   West, *Saloon*.

37   Etheredge, "Prohibition at 100"; Ogle, *Ambitious Brew*, 132.

38   Beienburg, *Prohibition, the Constitution, and States' Rights*.

5. Barley & Corn

1   Orsi, *Citizen Explorer*, 289.

2   Barth, *Chemistry of Beer*, 11–12.

3   Germanic peoples have been brewing with barley for thousands of years. Nelson, *Barbarian's Beverage*, 80.

4   For more on the struggles of feeding the Front Range and supply mining camps, see Carl Abbott, Stephen J. Leonard, and Thomas Noel, *Colorado: A History of the Centennial State*.

5   Mosher and Trantham, "Food" for the Brew.

6   Gupta, Abu-Ghannam, and Gallaghar, *Barley for Brewing*.

7   "Little Idaho a Money Saver," *Grangeville Globe*, January 11, 1912, image 5.

8   "Little Idaho a Money Saver."

9   "Electricity on Farm," *Yerington Times*, May 25, 1912, p. 3, image 3.

10   "Correspondence from the Colorado River," *Pioche Daily Record*, January 18, 1875, image 2.

11   "Leland Stanford: An Interesting Talk with the California Senator," *Helena Weekly Herald*, May 17, 1888, image 1.

12   Stika, "Early Iron Age."

13 For example, Max Nelson argues that ancient Egyptian baking and brewing can be seen as early as 2500 BCE. One confounding and perceptually interesting debate is which came first: bread or beer. Their formation, ingredients, and equipment are so similar that it is difficult to parse out exactly which was the purpose of identified archaeological sites. Nelson, *Barbarian's Beverage*; Valamoti, "Brewing Beer in Wine Country?"

14 For more on the scientific and mechanization of brewing in the late nineteenth and early twentieth centuries, see Walther et al., "Development of Brewing Science"; and Foda, *Egypt's Beer*.

15 Farber and Barth, *Mastering Brewing Science*, 2–3.

16 Lewis and Bamforth, "Raw Materials," 80–82.

17 Lewis and Bamforth, "Raw Materials," 80–82; Farber and Barth, *Mastering Brewing Science*, 2–3.

18 "All Imported Malt Is Not Alike."

19 "Van Blatz Beer," *Butte Daily Miner*, January 1, 1889.

20 "Trade Relations with Canada," *Salt Lake Herald*, May 6, 1891, p. 2, image 2.

21 Palen, "Protection, Federation and Union."

22 "Catching at a Straw," *Ketchum Keystone*, November 8, 1890, image 2.

23 "The Heavy Tax on Barley: Buffalo Interests to Be Ruthlessly Sacrificed; How the M'kinley Bill Will Cause Widespread Ruin—Comments of Canadian Newspapers," *New York Times (1857–1922)*, September 12, 1890, p. 5, https://search-proquest-com.aurarialibrary.idm .oclc.org/historical-newspapers/heavy-tax-on-barley/docview/ 94798113/se-2?accountid=14506.

24 "Rights of Buyer and Seller," *Salt Lake Herald*, October 7, 1894, p. 4, image 4.

25 Bettenhausen et al., "Influence of Malt Source."

26 Barley is just one of the agricultural products that Chicago brought into their fold. Most famously, William Cronon discusses this in *Nature's Metropolis*.

27 "One More for the McKinley Bill," *Fergus County Argus*, December 12, 1891.

28 "The Barley Question," *Weekly Tribune*, March 17, 1893.

29 Kok et al., "Brewing with Malted Barley."

30 "Corn and Chemicals," *Eddy Current*, February 6, 1897, image 4.

31 "Catching at a Straw," *Ketchum Keystone*, November 8, 1890, image 2.

32 Adulterations and stretching liquor supplies were significant issues in the Rocky Mountains, where supplies were limited. See, for example, West, *Saloon*; Noel, *City and the Saloon*, 6.

33 *Wadsworth Dispatch*, December 26, 1896, image 2. The veracity of these claims is unclear, but that did not deter them from making such accusations.

34 Barth, *Chemistry of Beer*, 17; Briggs et al., *Brewing*, 17.

35 For more on the scientific (rather than taste or perception) positives and negatives to adjuncts, see Lewis and Bamforth, "Raw Materials"; Bogdan and Kordialik-Bogacka, "Alternatives to Malt in Brewing"; Ceccaroni et al., "Specialty Rice Malt Optimization."

36 "A. Fisher Brewing Co.," *Salt Lake Herald*, August 18, 1888, p. 3, image 3.

37 "It's Good Beer," *Anaconda Standard*, February 7, 1897, p. 6, image 6.

6. Hops & Cones

1 United States Brewers' Association, *Year Book*, 198–99.

2 Peter Kopp is the expert in historic hop growing, particularly in Oregon's Willamette Valley. See Kopp, *Hoptopia*.

3 Schimke, "Brewing Terroir"; Bell, "Terroir of Beer."

4 Kopp, *Hoptopia*, 16.

5 Kopp, *Hoptopia*.

6 Neihart, "Frontier Beer," 35.

7 This is not to say hops do not grow here or that in recent years, efforts have not been made to produce them commercially. Simply, growing hops in the Rockies requires significantly more effort and care to produce an inferior product.

8 *Rocky Mountain News*, July 12, 1867, p. 2.

9 "Hop Culture," *White Pine News*, December 23, 1893.

10 "Hop Culture."

11 "Statistics of Agriculture," in *Report on the Statistics of Agriculture in the United States at the Tenth Census, 1890* (US Census Department, 1895), 106–14, https://www.census.gov/prod/www/decennial.html.

12 "Fourteenth Census of the United States: 1920—Number 568 Bulletins Statistics for the States and Its Counties," US Census

Department, 1919, https://www.census.gov/prod/www/decennial.html.

13 *Silver State*, August 15, 1901, image 1.

14 *Silver State*, August 15, 1901, image 1.

15 *Gem State Rural*, May 23, 1907, image 4.

16 *Silver State*, August 15, 1901, image 1.

17 "The Hop Trade," *Carson Daily Appeal*, from the *San Francisco Bulletin*, November 1, 1876, image 2.

18 Kopp, *Hoptopia*, 38.

19 "Selling at a Loss," *Caldwell Tribune*, June 22, 1907, image 11.

20 "Selling at a Loss."

21 "The Money Market," *Wood River Times*, September 6, 1888, image 2.

22 *Pioche Daily Record*, October 8, 1896, image 2.

23 "Stock and Crop Notes," *Clearwater Republican*, December 4, 1914, image 3.

24 "California Hops Are Selling Fast," *Parma Herald*, July 8, 1915, image 3.

25 "Boca Beer," *Pioche Daily Record*, August 20, 1876, image 2.

26 For example, *The Salt Lake Herald* had a column called "Bulls and Bears" that reported on a number of New York markets, hops being one of them.

27 "Hops Have Soared to 50 Cent Mark," *Tonopah Daily Bonanza*, August 23, 1911, image 3.

28 Kopp, *Hoptopia*, 3.

7. Local Flair

1 Joke lifted from "Joe Didn't Want It," *Central Nevadan*, June 20, 1895, image 3.

2 Ogle, *Ambitious Brew*.

3 Much has been written on the domination of lager in the US. For some easy references, see Ogle, *Ambitious Brew*.

4 *Daily Mining Journal*, September 15, 1864.

5 "Brewed in Pueblo," *Indicator*, vol. 10, no. 16, June 3, 1899.

6 "Brewed in Pueblo."

7 For example, Gaw's Brewery in Leadville, Colorado, lasted for decades and only went under after beers from outside the vicinity—such as

from Denver and further abroad—began appearing under vendors. I presented on this issue. Neihart, "Brewing Social Spaces."

8   "Fourth of July Story," *Florence Daily Tribune*, vol. 13, no. 107, July 7, 1904.

9   "Fourth of July Story."

10 "Car and Beer Wagon Struck; Spoiled Beer," *Salt Lake Herald-Republican*, August 21, 1910, sec. 1, p. 3, image 3.

11 *Cheyenne Daily Leader*, June 4, 1874, image 4.

12 "A Good Scheme," *Morning Appeal*, October 15, 1902, image 3.

13 "A Good Scheme," *Morning Appeal*, October 15, 1902, image 3.

14 All quotations from "About the City," *Anaconda Standard*, June 16, 1895, p. 3, image 3.

15 *Anaconda Standard*, April 1, 1899, p. 8, image 8.

16 *Idaho Springs Advance*, vol. 2, no. 85, September 28, 1882.

17 "A Big Institution," *Leader*, April 29, 1892; "The Phillip Zang Brewing Co.," advertisement in *Gunnison Tribune*, vol. 13, no. 23, October 22, 1892, p. 2, col. 3.

18 "Dillion Brewery One of the Pleasant Spots," *Butte Intermountain*, January 1, 1901, p. 6, image 16; "The City of Albuquerque: Metropolis of New Mexico—the Railroad, Manufacturing, and Commercial Center of the Southwest," *Albuquerque Daily Citizen*, October 13, 1902, image 12.

19 This is not the only shot between the two cities' beer wars but perhaps one of the more enjoyable ones. *Colorado Daily Chieftain*, March 22, 1873.

20 "Frozen Beer," *Colorado Daily Chieftain*, March 21, 1872.

21 *Los Cerrillos Rustler*, April 17, 1891, image 8.

22 "The San Juan Brewery," *San Juan Prospector*, May 3, 1884.

23 "San Juan Brewery."

24 Beer is in quotes, as some may dispute this constituting true beer. "A Pleasant Drink," *Cecil Whig*, June 25, 1859. For more on the malleability of beer, see Lars Marius Garshol, *Historical Brewing Techniques: The Lost Art of Farmhouse Brewing* (Brewers Publications, 2020).

25 Coors and Budweiser, like many other major corporations that court the same market, are often embroiled in litigation and legal battles. See Baur, "Anheuser-Busch."

26 Boulton, *Encyclopaedia of Brewing*, 594–95; Stubbs, "Captain Cook's Beer."

27 Tittensor, *Shades of Green*, 122.

28 Spruce beer was commonly made on the march. For more on the use by explorers and soldiers and its healthfulness, see Charters, "Diseases, Wilderness Warfare"; Janzen, "'Of Consequence to the Services'"; George, "Then & Now"; Barko, "French Garden at La Perouse"; Yagi Jr., "Surviving the Wilderness"; Tittensor, "Ships, Surveyors, Scurvy and Spruces"; and Lamb, May, and Harrison, "Enigma."

29 Prior, *Authentick Narrative*, 25.

30 *Publican's Own Book.*

31 "How to Make Spruce Beer," *Belmont Chronicle*, July 13, 1871, image 4.

32 "How to Make Spruce Beer," *Belmont Chronicle*.

33 "How to Make Spruce Beer," *Fayetteville Observer*, August 26, 1869, image 4; "How to Make Spruce Beer," *Dodgeville Chronicle*, July 17, 1868, image 4; "Summer Beverages," *West-Jersey Pioneer*, July 26, 1877, image 1; "Spruce Beer," *Diamond Drill*, October 31, 1908, image 2.

34 "Spruce Beer," *Diamond Drill*, October 31, 1908, image 2.

35 "Recipe for Spruce Beer," *Washington Herald*, January 5, 1913, image 28.

36 "Spruce Beer," *Salt Lake Herald*, June 14, 1896, image 17.

37 *The Bee*, June 28, 1877.

38 *Boise News*, October 20, 1863, image 3.

39 "How to Make Spruce Beer," *Belmont Chronicle*, July 13, 1871, image 4.

40 *Sierra County Advocate*, June 21, 1889, image 1.

41 *Sierra County Advocate*, June 21, 1889, image 1.

42 *Las Vegas Free Press*, March 3, 1892, image 2.

43 "The Differences of Opinion," *Eddy Current*, November 26, 1898, image 1.

8. Drinking Beer, Tasting Terroir

1 Farris and Reibstein, "How Prices."

2 Chowdhury, "Measuring the Relationship," 4.

3 "Big Bills for Booze," *Laramie Republican*, May 13, 1916, image 1.

4 J. E. Jones, "At National Capital," *Rathdrum Tribune*, July 11, 1913, image 1; "Beer Consumption Drops," *Clearwater Republican*, January 2, 1913, image 2.

5 "General Tables," US Census Bureau, 1901, p. 10; "Statistics of the Population," US Census Bureau, 1901, p. xx.
6 "Seven States Go into Dry Column," *Montpelier Examiner*, January 7, 1916, image 2.
7 "Seven States Go into Dry Column."
8 *Truth*, August 5, 1905, p. 8, image 8.
9 "Dream of the Mayor May Yet Be Realized," *Laramie Republican*, March 5, 1908, p. 7, image 7.
10 "Loophole Seen in 'Saving' Cause of the Dry Law," *Logan Republican*, July 28, 1917, p. 5, image 5.
11 "Back to the Ice Cream Soda Souse," *Payette Enterprise*, May 26, 1910, image 1.
12 "Denver Saloons Will Close," *Glenwood Post*, December 11, 1915, image 1.
13 "Some Reasons Why Idaho Will Vote for State Wide Prohibition," *Clearwater Republican*, April 10, 1914, image 8.
14 Mittal, "Bud Light Is Still Sinking."

9. Exclusions

1 Shah, "Beer Made from a Brewmaster's Beard."
2 Gordon Strong with Kristen England, "American Wild Ale," Beer Judge Certification Program 2021 Beer Style Guidelines, 2021, p. 6.
3 *Salt Lake Tribune*, May 20, 1909, p. 6, image 6.
4 "Cincinnati Boss Brewers Refuse to Increase Wages," *Denver Labor Bulletin*, February 26, 1916, image 1.
5 "Cincinnati Boss Brewers Refuse to Increase Wages."
6 "St. Louis Brewery Strike Is Ended," *Roswell Daily Record*, April 3, 1907, image 1.
7 "Brewery Strike," *Fergus County Democrat*, October 25, 1910, image 1.
8 "Brewery Strike Ends," *Salt Lake Tribune*, February 9, 1911, image 1.
9 "Unregulated Distilling of Liquor Would Result from State-Wide Prohibition," *Keota News*, September 25, 1914, image 6.
10 *Havre Herald*, August 31, 1906, p. 7, image 7.
11 *Havre Herald*, August 31, 1906, p. 7, image 7.
12 *Carson Daily Appeal*, May 4, 1909, image 2.

13 "Albuquerque to Have New $150,000 Brewery," *Albuquerque Daily Citizen*, October 4, 1907, p. 5, image 5.

14 "More New Buildings," *Anaconda Standard*, April 15, 1895, p. 3, image 3.

10. Nostalgia & Memory

1 Holtzman, "Food and Memory"; Townsend, "Thomas Reid."
2 Dighe, "Taste for Temperance," 17–48.
3 Dighe, "Taste for Temperance," 35.
4 See, for example, Verma and Rajendran, "Effect of Historical Nostalgia"; Li et al., "Effect of Nostalgia."
5 Chang and Feng, "Bygone Eras."
6 Pace, "Does Guinness Actually Taste Better." For more on why, scientifically, Guinness does taste better in Ireland, see Carey, "Scientific Evidence"; and Pomranz, "Does Guinness Really Taste Better."
7 Hart, "Drink Beer for Science."
8 Reed, "How Nostalgia Helped Kill," 35–36.
9 Schultz, "Pabst Bets on Beer Nostalgia."
10 Limerick, *Legacy of Conquest*, 24–28.
11 Crawley and Gabe, "Beercations," 193–205.
12 "Product Locator."
13 "About."
14 "Legend of the Tommyknockers."
15 "About Us."

11. Fermenting Terroir & Future Directions

1 Kuchar, "Fort Collins, Colorado"; Denver Microbrew Tour, "Why Denver."
2 "Rich History."

Bibliography

Periodicals

Albuquerque Daily Citizen
The American Brewer
The Anaconda Standard
The Bee
The Belmont Chronicle
Boise News
Boulder County News
Butte Daily Miner
Butte Intermountain
The Caldwell Tribune
The Carson Daily Appeal
The Cecil Whig
Central Nevadan
The Cheyenne Daily Leader
Clearwater Republican
Colorado Daily Chieftain
Daily Enterprise
The Daily Independent
Daily Inter Mountain
Daily Mining Journal
DeLamar Nugget
Delta Independent
The Denver Labor Bulletin
The Diamond Drill
Dillon Tribune
The Dodgeville Chronicle
The Eddy Current

The Eureka Sentinel
Evening Standard
The Fayetteville Observer
Fergus County Argus
Fergus County Democrat
Florence Daily Tribune
Gem State Rural
Glenwood Post
Goodwin's Weekly
The Grangeville Globe
The Greeley Tribune
Gunnison Tribune
The Havre Herald
Helena Weekly Herald
Idaho Springs Advance
The Indicator
The Keota News
The Ketchum Keystone
La Junta Tribune
Laramie Republican
Las Vegas Daily Gazette
Las Vegas Free Press
The Leader
Lewiston Evening Teller
Livingston Enterprise
Logan Republican
Los Cerrillos Rustler
Montpelier Examiner
Morning Appeal
Neihart Herald
The New Era
The New York Times
The Ogden Standard
The Parma Herald
Payette Enterprise
Pioche Daily Record
The Rathdrum Tribune

Rocky Mountain News
Roswell Daily Record
The Salt Lake Herald
The Salt Lake Herald-Republican
The Salt Lake Tribune
San Francisco Bulletin
The San Juan Prospector
Santa Fe Daily New Mexican
Sierra County Advocate
The Silver State
Tonopah Bonanza
Tonopah Daily Bonanza
Truth: The Western Weekly
The Wadsworth Dispatch
Walker Lake Bulletin
The Washington Herald
The Weekly Tribune
The Western Brewer and Journal of the Barley, Malt and Hop Trades
West-Jersey Pioneer
The White Pine News
Wood River Times
Yerington Times

Primary Sources

Account of [[Beer]] and Rum Sold to the Earl of Bellomont. Business and financial document. The Gilder Lehrman Institute of American History, GLC03107.02049. Available through Adam Matthew, Marlborough, American History, 1493–1945. Accessed February 20, 2022. http://www.americanhistory.amdigital.co.uk.du.idm.oclc.org/Documents/Details/GLC03107.02049.

Livingston, Robert. *Account of Provisions and Beer Delivered to the Forces at Albany*. Business and financial document. The Gilder Lehrman Institute of American History, GLC03107.00144. Available through Adam Matthew, Marlborough, American History, 1493–1945. Accessed February 20, 2022. http://www.americanhistory.amdigital.co.uk.du.idm.oclc.org/Documents/Details/GLC03107.00144.

Prior, Thomas. *The Authentick Narrative of the Success of Tar Water, in Curing a Great Number and Variety of Distempers; with Remarks: By Thomas Prior, Esq; Carefully Abridged. To Which Are Subjoined, Two Letters from the Author of Siris: Shewing the Medicinal Properties of Tarwater, and the Best Manner of Making It.* London, printed 1746. Boston, NE, reprinted and sold by Rogers and Fowle in Queen-Street, 1749.

The Publican's Own Book and Domestic Brewer's Guide: Containing Full Information for Choosing Water, Malt and Hops . . . Brewing Cottage Beer . . . Shropshire, Burton, and Scotch Ales, Porter . . . Also, Copious Instructions in British Wine Making, Management of Spirits, Making Cordials, &c. Printed and published by G. Bateman, 1840.

Rush, Benjamin. *An Enquiry into the Effects of Spirituous Liquors upon the Human Body: And Their Influence upon the Happiness of Society.* Philadelphia, printed by Thomas Bradford, 1784.

Van Horne, Johanna Livingston. *Johanna Livingston to Robert Livingston re: Receipt of Spoiled Beer.* Available through Adam Matthew, Marlborough, American History, 1493–1945. Accessed February 21, 2022. http://www.americanhistory.amdigital.co.uk.du.idm.oclc.org/Documents/Details/GLC03107.00958.

Van Horne, Johanna Livingston. *Johanna Livingston to Robert Livingston re: Request for Livingston to Send Beer and Biskett.* Correspondence. The Gilder Lehrman Institute of American History, GLC03107.00960. Available through Adam Matthew, Marlborough, American History, 1493–1945. Accessed February 21, 2022. http://www.americanhistory.amdigital.co.uk.du.idm.oclc.org/Documents/Details/GLC03107.00960.

Websites

"About." Grand Teton Brewing. Accessed July 30, 2021. http://grandtetonbrewing.com/about-2/.

"About Us." Bow and Arrow Brewing. Accessed March 21, 2022. https://www.bowandarrowbrewing.com.

"Legend of the Tommyknockers." Tommyknocker Brewery. Accessed July 30, 2021. https://tommyknocker.com.

"Product Locator." Molson Coors Beverage Company. Accessed April 18, 2021. https://www.coloradonative.com/locator.

"A Rich History." 10th Mountain Whiskey & Spirit. Accessed March 4, 2023. https://10thwhiskey.com/pages/our-story.

US Census Bureau. https://www.census.gov/.

Secondary Sources

Alberts, Brian. "Brewed to Stay: Brewed Culture, Ethnicity, and the Market Revolution." PhD diss., Purdue University, 2018. Proquest.

"All Imported Malt Is Not Alike." *American Brewer* vol. 68, no. 4 (1935): 11.

Alworth, Jeff. *The Beer Bible.* New York: Workman, 2015.

Andrews, Thomas G. *Coyote Valley: Deep History in the High Rockies.* Cambridge, MA: Harvard University Press, 2015.

Appel, Susan K. "Artificial Refrigeration and the Architecture of 19th-Century American Breweries." *Journal of the Society for Industrial Archeology* 16, no. 1 (1990): 21.

Barko, Ivan. "The French Garden at La Perouse." *Australian Garden History* 24, no 2 (Winter 2012): 9.

Barth, Roger. *The Chemistry of Beer: The Science in the Suds.* John Wiley & Sons, 2013.

Bat-Oyun, Tserenpurev, Baasandai Erdenetsetseg, Masato Shinoda, Takahiro Ozaki, and Yuki Morinaga. "Who Is Making 'Airag' (Fermented Mares' Milk)? A Nationwide Survey of Traditional Food in Mongolia." *Nomadic Peoples* 19, no. 1 (2015): 7–29. http://www.jstor.org/stable/24772842.

Baum, Dan. *Citizen Coors: A Grand Family Saga of Business, Politics, and Beer.* New York: Harper Paperbacks, 2001.

Baur, Scott. "Anheuser-Busch Wins Latest Round of Beer Wars Against Molson." Associated Press, May 1, 2020.

Beienburg, Sean. *Prohibition, the Constitution, and States' Rights.* Chicago: University of Chicago Press, 2019.

Bell, Emma. "The Terroir of Beer." *Edible Toronto*, October 11, 2018. https://edibletoronto.ediblecommunities.com/drink/terroir-beer (page no longer extant).

Bennett, Judith. *Ale, Beer, and Brewsters in England: Women's Work in a Changing World, 1300–1600.* Oxford: Oxford University Press, 1999.

Bettenhausen, Harmonie M., et al. "Influence of Malt Source on Beer Chemistry, Flavor, and Flavor Stability." *Food Research International* 113 (2018): 487–504. https://doi.org/10.1016/j.foodres.2018.07.024.

Bodnár, V., K. Nagy, Á. Cziboly, and G. Bárdos. "Alcohol and Placebo: The Role of Expectations and Social Influence." *International Journal of Mental Health and Addiction* 19 (2021): 2292–305. https://doi.org/10.1007/s11469-020-00321-0.

Bogdan, Paulina, and Edyta Kordialik-Bogacka. "Alternatives to Malt in Brewing." *Trends in Food Science & Technology* 65 (2017): 1–9. https://doi.org/10.1016/j.tifs.2017.05.001.

Boulton, Christopher M., ed. *Encyclopaedia of Brewing*. Hoboken, NJ: John Wiley & Sons, 2013.

Bramwell, Lincoln. *Wilderburbs: Communities on Nature's Edge*. Seattle: University of Washington Press, 2014.

Briggs, Dennis E., Chris A. Boulton, Peter A. Brookes, and Roger Stevens, eds. *Brewing: Science and Practice*. Woodhead Publishing Series in Food Science, Technology and Nutrition. Woodhead, 2004.

Broderick, Amanda J., and Rene D. Mueller. "A Theoretical and Empirical Exegesis of the Consumer Involvement Construct: The Psychology of the Food Shopper." *Journal of Marketing Theory and Practice* 7, no. 4 (1999): 97–108. http://www.jstor.org/stable/23232704.

Cabras, Ignazio, and David M. Higgins, eds. *The History of the Beer and Brewing Industry*. New York: Routledge, 2018.

Carey, Bjorn. "Scientific Evidence: Guinness Tastes Better in Ireland." Live Science, March 17, 2011. https://www.livescience.com/33125-guinness-tastes-better-in-ireland.html.

Chang, Chun-Tuan, and Ching-Chiao Feng. "Bygone Eras vs. the Good Ol' Days: How Consumption Context and Self-Construal Influence Nostalgic Appeal Selection." *International Journal of Advertising* 35, no. 3 (2016): 589–615.

Charters, Erica M. "Diseases, Wilderness Warfare, and Imperial Relations: The Battle for Quebec, 1759–1790." *War in History* 16, no 1 (January 2009): 1–24.

Childers, Michael. *Colorado Powderkeg: Ski Resorts and the Environmental Movement*. Lawrence: University Press of Kansas, 2012.

Chowdhury, Sultana Razia. "Measuring the Relationship Among the Advertisement Expenditure, Sales Revenue, and Profit on Steel Industries and Banking Industries in Bangladesh," *European Journal of Business and Management* 9, no. 9 (2017).

Conway, Jan. "Global Beer Production 1998–2020." Statista, October 15, 2021. https://www.statista.com/statistics/270275/worldwide-beer-production/.

Crawley, Andrew, and Todd Gabe. "Beercations: A Spatial Analysis of the U.S. Craft Brewery and Tourism Sectors." In *Agritourism, Wine Tourism, and Craft Beer Tourism: Local Responses to Peripherality Through Tourism Niches*, edited by Marla Guilia Pezzi, Alessandra Faggian, and Niel Reid. Routledge, 2021.

Cronon, William. *Nature's Metropolis: Chicago and the Great West*. 1st ed. New York: W. W. Norton, 1991.

Cutler, Hugh C., and Martín Cárdenas. "Chicha, a Native South American Beer." *Botanical Museum Leaflets, Harvard University* 13, no. 3 (1947): 33–60.

Dayana Ceccaroni, Valeria Sileoni, Ombretta Marconi, Giovanni De Francesco, Eung Gwan Lee, and Giuseppe Perretti. "Specialty Rice Malt Optimization and Improvement of Rice Malt Beer Aspect and Aroma." *LWT—Food, Science, and Technology* 99 (2019): 299–305. https://doi.org/10.1016/j.lwt.2018.09.060.

De Keersmaecker, Jacques. "The Mystery of Lambic Beer." *Scientific American* 275, no. 2. (August 1996): 74–80.

Denver Microbrew Tour. "Why Denver Is Still the Napa Valley of Beer." https://www.denvermicrobrewtour.com/why-denver-is-still-the-napa-valley-of-beer/.

Dighe, Ranjit S. "A Taste for Temperance: How American Beer Got to Be So Bland." In Cabras and Higgins, *History of the Beer and Brewing Industry*, 2017.

Dougherty, Percy H. *The Geography of Wine: Regions, Terroir and Techniques*. New York: Springer, 2012.

Dutton, Jacqueline, and Peter J. Howland, eds. *Wine, Terroir and Utopia: Making New Worlds*. New York: Routledge, 2020.

Encyclopaedia Britannica. "Laramide Orogeny." Accessed March 12, 2021. https://www.britannica.com/science/Laramide-orogeny.

Etheredge, Stacy. "Prohibition at 100: A Comprehensive and Annotated Federal Legislative History." *Legal Reference Services Quarterly* 39, nos. 2–3 (2020): 204–42.

Eumann, M., and C. Schaeberle. "Water." In *Brewing Materials and Processes*, edited by Charles W. Bamforth. Academic Press, 2016.

Farber, Matthew, and Roger Barth. *Mastering Brewing Science: Quality and Production.* Hoboken, NJ: Wiley, 2019.

Farris, Paul, and David J. Reibstein. "How Prices, Ad Expenditures, and Profits Are Linked." *Harvard Business Review,* November 1979.

Fiege, Mark. *The Republic of Nature: An Environmental History of the United States.* Seattle: University of Washington Press, 2012.

Flavin, Susan, et al. "An Interdisciplinary Approach to Historic Diet and Foodways: The FoodCult Project." *European Journal of Food Drink and Society* 1, no. 1 (2021): article 3.

Foda, Omar D. *Egypt's Beer: Stella, Identity, and the Modern State.* 1st ed. Austin: University of Texas Press, 2019.

Gabel, B. "Wine Origin Authentication Linked to Terroir—Wine Fingerprint." *BIO Web of Conferences* 15 (2019): 2033.

Gammons, Christopher H., John J. Metesh, and Terence E. Duaime. "An Overview of the Mining History and Geology of Butte, Montana." *Mine Water and the Environment* 25, no. 2 (2006): 70–75.

Gatrell J., D. Nemeth, and C. Yeager. "Sweetwater, Mountain Springs, and Great Lakes: A Hydro-Geography of Beer Brands." In Patterson and Hoalst-Pullen, *Geography of Beer.*

George, Emily. "Then & Now: The Evolution of Army Rations." *On Point* 17, no. 4 (Spring 2012): 67.

Gomez, Rocio. *Silver Veins, Dusty Lungs: Mining, Water, and Public Health in Zacatecas, 1835–1946.* Lincoln: University of Nebraska Press, 2020.

Gupta, M., N. Abu-Ghannam, and E. Gallaghar. "Barley for Brewing: Characteristic Changes During Malting, Brewing and Applications of Its by-Products." *Comprehensive Reviews in Food Science and Food Safety* 9 (2010): 318–28.

Hämäläinen, Pekka. *The Comanche Empire.* New Haven: Yale University Press, 2008.

Harding, Graham. "Inventing Tradition and Terroir: The Case of Champagne in the Late Nineteenth Century." In Dutton and Howland, *Wine, Terroir and Utopia.*

Harry, John. *Stevens Point Brewing Company.* Arcadia Publishing, 2019.

Hart, Jarrett. "Drink Beer for Science: An Experiment on Consumer Preferences for Local Craft Beer." *Journal of Wine Economics* 13, no. 4 (2018): 429–41.

Harvey, Matt, Leanne White, and Warwick Frost. eds. *Wine and Identity: Branding, Heritage, Terroir*. New York: Routledge, 2014.

Hoalst-Pullen, Nancy, and Mark W. Patterson. *The Geography of Beer: Regions, Environment, and Societies*. Springer, 2014.

Hogan, John, and Judy Brady. *The Great Chicago Beer Riot: How Lager Struck a Blow for Liberty*. History Press Library Editions, 2015.

Holtzman, Jon D. "Food and Memory." *Annual Review of Anthropology* 35 (2006): 361–78. http://www.jstor.org/stable/25064929.

Hoverson, Doug. *The Drink That Made Wisconsin Famous: Beer and Brewing in the Badger State*. Minneapolis: University of Minnesota Press, 2019.

Hoverson, Doug. *Land of Amber Waters: The History of Brewing in Minnesota*. Minneapolis: University of Minnesota Press, 2007.

Howland, Peter J., and Jacqueline Dutton. "Making New Worlds: The Utopian Potentials of Wine and Terroir." In *Wine, Terroir and Utopia*, 6–7.

Janzen, Olaf U. "'Of Consequence to the Services': The Rationale Behind Cartographic Surveys in Early Eighteenth-Century Newfoundland." In *War and Trade in Eighteenth-Century Newfoundland*. Liverpool University Press, 2013.

Knoedelseder, William. *Bitter Brew: The Rise and Fall of Anheuser-Busch and America's Kings of Beer*. Harper Business, 2012.

Kok, Y. J., et al. "Brewing with Malted Barley or Raw Barley: What Makes the Difference in the Processes?" *Applied Microbiology and Biotechnology* 103 (2019): 1059–67.

Kopp, Peter. *Hoptopia: A World of Agriculture and Beer in Oregon's Willamette Valley*. Oakland: University of California Press, 2016.

Kuchar, Kristen. "Fort Collins, Colorado: The Napa Valley of Beer." Why Wait to See the World, February 10, 2013.

Lamb, Jonathan, James May, and Fiona Harrison. "Enigma." In *Scurvy: The Disease of Discovery*. Princeton University Press, 2017.

Lears, T. J. Jackson. *Rebirth of a Nation: The Making of Modern America, 1877–1920*. 1st ed. New York: Harper, 2009.

Lewis, M. J., and C. W. Bamforth. "Raw Materials." In *Essays in Brewing Science*. Boston: Springer, 2006. https://doi-org.aurarialibrary.idm.oclc.org/10.1007/0-387-33011-9_8.

Libkind, Diego, et al. "Microbe Domestication and the Identification of the Wild Genetic Stock of Lager-Brewing Yeast." *Proceedings of the National Academy of Sciences of the United States of America* 108, no. 35 (2011): 14539–44. http://www.jstor.org.aurarialibrary.idm.oclc.org/stable/27979327.

Limerick, Patricia Nelson. *The Legacy of Conquest: The Unbroken Past of the American West.* New York: Norton, 1988.

Limerick, Patricia Nelson. *Something in the Soil: Legacies and Reckonings in the New West.* New York: W. W. Norton, 2000.

Limerick, Patricia Nelson, Jason L. Hanson, and Timothy Brown. *A Ditch in Time: The City, the West, and Water.* Golden, CO: Fulcrum, 2012.

Linderberger, Hudson. "Despite the Pandemic, Super Bowl Beer Sales Look to Score." Forbes, February 2, 2021.

Logan, Amanda L., Christine A. Hastorf, and Deborah M. Pearsall. "'Let's Drink Together': Early Ceremonial Use of Maize in the Titicaca Basin." *Latin American Antiquity* 23, no. 3 (2012): 235–58.

MacDonnell, Lawrence J., and University of Colorado Boulder. *From Reclamation to Sustainability: Water, Agriculture, and the Environment in the American West.* Natural Resources Law Center. Niwot: University Press of Colorado, 1999.

Mayer, Aaron Jay, Matthew Sayre, and Justin Jennings. "Coming Together to Toast and Feed the Dead in the Cotahuasi Valley of Peru." *Ethnobiology Letters* 8, no. 1 (2017): 46–53.

McGovern, Patrick. *Ancient Brews: Rediscovered and Re-Created.* New York: Norton, 2017.

McNeill, John Robert, and George Vrtis. *Mining North America: An Environmental History Since 1522.* Oakland: University of California Press, 2017.

Melewar, T. C., and Heather Skinner. "Territorial Brand Management: Beer, Authenticity, and Sense of Place." *Journal of Business Research* 116 (2020): 680–89.

Mittal, Vikas. "Bud Light Is Still Sinking: Here's Why It Really Lost Its Crown." *The Hill*, October 21, 2023.

Moore, Jerry D. "Pre-Hispanic Beer in Coastal Peru: Technology and Social Context of Prehistoric Production." *American Anthropologist* 91, no. 3 (1989): 682–95.

Mosher, M., and K. Trantham. *The "Food" for the Brew.* In *Brewing Science: A Multidisciplinary Approach.* Cham: Springer, 2017. https://doi-org .aurarialibrary.idm.oclc.org/10.1007/978-3-319-46394-0_5.

National Park Service. "Weather." Accessed March 12, 2021. https://www .nps.gov/grba/planyourvisit/weather.htm.

Neihart, Braden. "Beer Wars: Capitalism and Conflict in Rock Island, IL. 1880–1900." *Journal of Illinois State History* 115, nos. 2–3 (Summer 2022): 132–48.

Neihart, Braden. "Brewing Social Spaces: Beer and Business in Leadville, 1879–1905." WHA Annual Conference, 2020.

Neihart, Braden. "Frontier Beer: A Spatial Analysis of Denver Breweries, 1859–1876." MA thesis, Colorado State University, 2019. ProQuest.

Nelson, Max. *The Barbarian's Beverage: A History of Beer in Ancient Europe.* New York: Routledge, 2005.

Noel, Thomas J. *The City and the Saloon: Denver, 1858–1916.* Niwot: University Press of Colorado, 1996.

Ogle, Maureen. *Ambitious Brew: The Story of American Beer.* 1st ed. Orlando: Harcourt, 2006.

Orsi, Jared. *Citizen Explorer: The Life of Zebulon Pike.* New York: Oxford University Press, 2014.

Pace, Gina. "Does Guinness Actually Taste Better in Ireland?" *The Thrillist,* March 6, 2018. https://www.thrillist.com/drink/does-guinness-taste -better-in-ireland.

Palen, Marc-William. "Protection, Federation and Union: The Global Impact of the McKinley Tariff upon the British Empire, 1890–94." *Journal of Imperial and Commonwealth History* 38, no. 3 (2010): 395–418.

Parker, Bradley J., and Weston McCool. "Indices of Household Maize Beer Production in the Andes: An Ethnoarchaeological Investigation." *Journal of Anthropological Research* 71, no. 3 (2015): 359–400.

Parry, William T. "Geology and Utah's Mineral Treasures." In *From the Ground Up: A History of Mining in Utah,* edited by Colleen Whitley, 3–36. University Press of Colorado, 2006.

Philpott, William. *Vacationland: Tourism and Environment in the Colorado High Country.* Seattle: University of Washington Press, 2013.

Pomranz, Mike. "Does Guinness Really Taste Better in Ireland?" *Food & Wine,* August 25, 2016. https://web.archive.org/www.foodandwine .com/drinks/does-guinness-taste-better-ireland.

Pomranz, Mike. "125-Year-Old Olympia Beer Is Being Discontinued (at Least for Now)." *Food & Wine*, January 27, 2021. https://www.foodandwine.com/news/olympia-beer-ceases-production.

Price, Martin F., Alton C. Byers, Donald A. Friend, Thomas Kohler, and Larry W. Price, eds. *Mountain Geography: Physical and Human Dimensions*. University of California Press, 2013.

Reed, Charley. "How Nostalgia Helped Kill a Midwest Beer Brand's Revival." In *Beer Culture and Theory*.

Robins, Nicholas A. *Mercury, Mining, and Empire: The Human and Ecological Cost of Colonial Silver Mining in the Andes*. Bloomington: Indiana University Press, 2011.

Rodrigo, S., S. D. Young, M. I. Talaverano, and M. R. Broadley. "The Influence of Style and Origin on Mineral Composition of Beers Retailing in the UK." *European Food Research and Technology = Zeitschrift Für Lebensmittel-Untersuchung Und-Forschung. A* 243, no. 6 (June 2017): 931–39.

Rorabaugh, W. J. *The Alcoholic Republic, an American Tradition*. New York: Oxford University Press, 1979.

Schimke, Cedar. "Brewing Terroir: Unearthing the Distinct Regional Flavor of Hops." *The Growler*, February 26, 2018. https://www.growlermag.com/brewing-terroir-unearthing-distinct-regional-flavors-in-hops/.

Schultz, E. J. "Pabst Bets on Beer Nostalgia: Brewer Hopes for Resurgence of Classics Jax, Stroh's, Old Style." *Advertising Age* 87, no. 16 (August 22, 2016).

Shah, Khushbu. "Beer Made from a Brewmaster's Beard Yeast Is the Official Drink of No-Shave November." Eater, September 28, 2015. https://www.eater.com/2015/9/28/9409427/beer-made-from-mans-beard-yeast-rogue-ales-no-shave-november.

Smeets, Paul A. M., and Cees de Graaf. "Brain Responses to Anticipation and Consumption of Beer with and Without Alcohol." *Chemical Senses* 44, no. 1 (January 2019): 51–60. https://doi.org/10.1093/chemse/bjy071.

Smith, Andrew. *Fifteen Turning Points in the Making of American Beverages*. Columbia University Press, 2–4.

Smith, Duane A. *The Trail of Gold and Silver: Mining in Colorado, 1859–2009*. Boulder: University Press of Colorado, 2009.

Smith, Gregg. *Beer in America: The Early Years 1587–1840; Beer's Role in the Settling of America and the Birth of a Nation*. Boulder, CO: Brewer's Publications, 1998.

Stika, H. P. "Early Iron Age and Late Mediaeval Malt Finds from Germany—Attempts at Reconstruction of Early Celtic Brewing and the Taste of Celtic Beer." *Archaeological and Anthropological Sciences* 3 (2011): 41–48.

Stubbs B. J. "Captain Cook's Beer: The Antiscorbutic Use of Malt and Beer in Late 18th Century Sea Voyages," *Asia Pacific Journal of Clinical Nutrition* 12, no. 2 (2003):129–37.

Summitt, April R. *Contested Waters: An Environmental History of the Colorado River*. Boulder: University Press of Colorado, 2013.

Swinburn, Robert. "Deep Terroir as Utopia: Explorations of Place and County in Southeastern Australia." In Dutton and Howland, *Wine, Terroir and Utopia*, 222.

Tittensor, Ruth. *Shades of Green: An Environmental and Cultural History of the Sitka Spruce*. Oxbow Books, Windgather Press, 2016.

Tittensor, Ruth. "Ships, Surveyors, Scurvy and Spruces." In *Shades of Green*.

Tomasi, Diego, Federica Gaiotti, and Gregory V. Jones. *The Power of the Terroir: The Case Study of Prosecco Wine*. Basel: Springer, 2013.

Townsend, Dabney. "Thomas Reid and the Theory of Taste." *Journal of Aesthetics and Art Criticism* 61, no. 4 (2003): 341–51. http://www.jstor.org/stable/1559069.

Unger, Richard W. *Beer in the Middle Ages and the Renaissance*. Philadelphia: University of Pennsylvania Press, 2004.

Unger, Richard W. *A History of Brewing in Holland, 900–1900: Economy, Technology, and the State*. Leiden: Brill, 2001.

United States Brewers' Association. *The Year Book of the United States Brewers' Association*. New York: The Association, 1909.

Unwin, Tim. "*Terroir*: At the Heart of Geography." In Dougherty, *Geography of Wine*.

Valamoti, S. M. "Brewing Beer in Wine Country? First Archaeobotanical Indications for Beer Making in Early and Middle Bronze Age Greece." *Vegetation History and Archaeobotany* 27 (2018): 611–25.

Van Holle, A., et al. "The Brewing Value of Amarillo Hops (*Humulus Lupulus* L.) Grown in Northwestern USA: A Preliminary Study of Terroir Significance." *Journal of the Institute of Brewing* 123 (2017): 312–18.

Van Holle, A., et al. "Relevance of Hop Terroir for Beer Flavour." *Journal of the Institute of Brewing* 127, no. 3 (2021).

Verma, Anil, and G. Rajendran. "The Effect of Historical Nostalgia on Tourists' Destination Loyalty Intention: An Empirical Study of the World Cultural Heritage Site—Mahabalipuram, India." *Asia Pacific Journal of Tourism Research* 22, no. 9 (2017): 977–90. https://doi.org/10.1080/10941665.2017.1357639.

Walther, Andrea, Davide Ravasio, Fen Qin, Jürgen Wendland, and Sebastian Meier. "Development of Brewing Science in (and Since) the Late 19th Century: Molecular Profiles of 110–130 Year Old Beers." *Food Chemistry* 183 (2015): 227–34. https://doi.org/10.1016/j.foodchem.2015.03.051.

West, Elliott. *The Contested Plains: Indians, Goldseekers, & the Rush to Colorado.* Lawrence: University Press of Kansas, 1998.

West, Elliott. *The Saloon on the Rocky Mountain Mining Frontier.* Lincoln: University of Nebraska Press, 1979.

Western Regional Climate Center. "Average Statewide Precipitation for Western U.S. States." Accessed March 12, 2021. https://wrcc.dri.edu/Climate/comp_table_show.php?stype=ppt_avg.

Yagi, George, Jr. "Surviving the Wilderness: The Diet of the British Army and the Struggle for Canada, 1754–1760." *Journal of the Society for Army Historical Research* 89, no. 357 (Spring 2011): 66–86.

Yizhi, Li, Can Lu, Vanja Bogicevic, and Milos Bujisic. "The Effect of Nostalgia on Hotel Brand Attachment." *International Journal of Contemporary Hospitality Management* 31, no. 2 (2019): 691–717.

Yool, S., and A. Comrie. "A Taste of Place: Environmental Geographies of the Classic Beer Styles." In Patterson and Hoalst-Pullen, *Geography of Beer.*

Zhou, Jia, et al. "Wine Terroir and the Soil Bacteria: An Amplicon Sequencing-Based Assessment of the Barossa Valley and Its Sub-Regions." *Frontiers in Microbiology* 11 (2020–21): 597944.

Index

www.ingramcontent.com/pod-product-compliance
Lightning Source LLC
Chambersburg PA
CBHW071130280326
41935CB00010B/1172